Business Models for Sustainable Telecoms Growth in Developing Economies

Business Models for Sustainable Telecoms Growth in Developing Economies

Sanjay Kaul
Ericsson South Africa (PTY) Ltd, South Africa

Fuaad Ali
Habibi Enterprises International, South Africa

Subramaniam Janakiram
World Bank, USA

Bengt Wattenström
Ericsson, Sweden

John Wiley & Sons, Ltd

Other Wiley Editorial Offices

John Wiley & Sons Inc., 111 River Street, Hoboken, NJ 07030, USA

Jossey-Bass, 989 Market Street, San Francisco, CA 94103-1741, USA

Wiley-VCH Verlag GmbH, Boschstr. 12, D-69469 Weinheim, Germany

John Wiley & Sons Australia Ltd, 42 McDougall Street, Milton, Queensland 4064, Australia

John Wiley & Sons (Asia) Pte Ltd, 2 Clementi Loop #02-01, Jin Xing Distripark, Singapore 129809

John Wiley & Sons Canada Ltd, 6045 Freemont Blvd, Mississauga, ONT, L5R 4J3, Canada

Wiley also publishes its books in a variety of electronic formats. Some content that appears in print may not be available in electronic books.

Library of Congress Cataloging-in-Publication Data

Kaul, Sanjay, 1960–
    Business models for sustainable telecoms growth in developing economies / Sanjay Kaul, Fuaad Ali, Subramaniam Janakiram.
        p.   cm.
    Includes index.
    ISBN 978-0-470-51972-1 (cloth)
    1. Telecommunication–Developing countries.   2. Sustainable development–Developing countries.   I. Ali, Fuaad.   II. Janakiram, Subramaniam.   III. Title.
    HE8635.K38   2008
        384'.041091724 – dc22

                                                                    2007049349

British Library Cataloguing in Publication Data

A catalogue record for this book is available from the British Library

ISBN 978-0-470-51972-1 (HB)

Typeset by SNP Best-set Typesetter Ltd., Hong Kong.
Printed and bound in Great Britain by TJ International, Padstow, Cornwall.

Contents

This book is dedicated to all those who despite all the advances in communication technologies still do not have access or cannot afford any form of telecommunications.

Foreword

The contribution of communications networks to the development of nations can be likened to that of other key infrastructures such as roads, ports and railways. All stimulate trade, create jobs and generate wealth. Metcalf's law states that each new connection has an exponential impact. A 10% increase in mobile penetration, for example, can boost GDP by 1.2% in a typical emerging market and the effect on social capital is equally significant as families and friends connect more closely.

Mobile communication is not just about the voice. Some 7 billion text messages are sent every day, delivering a rich suite of services such as market pricing, health and security solutions. For the many millions of people that have previously been ineligible, financial services will increasingly be accessed via mobile phones.

Twenty years ago, the foundations of the mobile revolution were laid when 15 operators in 13 countries signed the Global System for Mobile Communications (GSM) Memorandum of Understanding (MOU). The MOU ensured that mobile phones would work across borders and that spectrum use would be harmonised within ITU guidelines. This combination has driven massive economies of scale. In 2007, more than 718 operators in 220 countries are signatories of the MOU and more than 1 billion phones will be sold, many for as little as US$25. With very low prepay top-up denominations, along with shared access solutions, mobile services are now affordable for bottom-of-the-pyramid consumers. Of today's 3 billion mobile subscribers, more than 65% are in emerging markets and this percentage will grow as we approach the 4 billion subscriber milestone in 2010.

The mobile revolution continues at an astounding pace. Since 2002, more than US$230 billion has been invested by GSM operators, extending the mobile footprint to more than 80% of the worlds population. Access will increase to well over 95% of the population by 2012 as investment ramps up and countries continue to liberalise their markets. Operators in sub-Saharan Africa, for example, invested US$35 billion between 1994 and 2006 but plan to invest US$50 billion between 2007 and 2012, an annualised average increase of 330%. Future investment will not only extend network coverage and capacity but also roll out mobile Internet and broadband data services using GPRS, EDGE and 3G HSPA technologies.

This book sets out to demonstrate how the private sector will meet the public sector's goals and connect the world in a sustainable way. The biggest threat to the mobile revolution is if governments over-regulate the industry and over-tax consumers. Some 20 countries still impose specific taxes on mobile communications that hike the price of calls.

It is fitting that the authors bring considerable experience from their work with Ericsson and the World Bank. Both the organizations have played an enormous role in the industry's success to date. Ericsson has been a pioneer in emerging markets, setting up shop many generations ago. The World Bank, through its IFC arm, financed and nurtured many nascent emerging market operators into leading global players. Emerging markets are driving innovation and industry growth. This book explains the business models that will sustain the mobile revolution and deliver universal access to voice and data services.

Gabriel Solomon
Senior Vice President
GSMA

Preface

So, why have we written this book? Today, a little more than 3 billion people have a mobile phone. Of the remaining 4 billion, some 48% live close to the poverty line, earning less than US$5 per day. These are the very people that have been starved of information because they lack basic access to communications. This book is written around understanding how telecommunications operators, service providers, telecoms regulators, governments and others can reach out to this so-called mass market. At the same time we try to answer the question of how communications providers can reach out to these markets and still continue to grow while maintaining profitability. We also attempt to show the important role that governments and regulators alike should play to promote communications access in mass markets within developing economies.

This book is targeted at communications industry leaders and c-level executives, regulators, governments, international organisations, advocacy groups, telecoms business analysts, donor organisations, educational institutions and other non-profit organisations, as well as the various community leaders in developing markets. The rationale for this is that leaders have a responsibility to ensure that no one is left out of the information society.

Some of the benefits of reading this book are to be found amongst the following:

- Obtaining a clear understanding of the current state of the telecommunications sector and identifying the inherent drivers and barriers to the global adoption of mobile services
- First-hand understanding directly from CEOs – presentation of results of multiple interviews and one-on-one discussions held with top-level players and most influential persons leading the telecommunications industries helping to bridge the digital divide

- An exposé of state-of-the-art frameworks that aim to evolve current business models to new ones that will provide sustainable growth for telecommunications in developing markets.

The final aims of this book are to stimulate the strategic thinking process amongst key stakeholders and the communities within which they operate in the telecoms industry and those working towards a fairer, more balanced society, and to provide concrete tools that can be used to ensure that growth in telecommunications becomes sustainable and can be taken to the 'real' mass market: the 4 billion people who do not have the means or access to mobile telecoms.

AIM, IMPORTANCE AND PURPOSE OF THIS BOOK

The central aim and purpose of this book is to determine the opportunities that exist in mass markets in developing economies and to provide insight into the new business models that exist for creating sustainable growth in these markets. This book is important to anyone that has an interest in establishing how communications can be taken to the mass market in developing countries. It is especially beneficial to people working in the telecommunications sector, students studying telecommunications engineering, donor governments, business, regulatory bodies, communications suppliers and non-government organisations (NGOs). The purpose of this book is to stimulate debate around key issues in telecommunications and to explore some sustainable business models that can be used to increase telecommunications penetration and promote economic growth in developing countries.

STRUCTURE FOR THE REST OF THE BOOK

Chapter 1: The Impact of Communications on Developing Markets

In this first chapter, we strive to set the stage for what follows in the rest of this book. We review mobile infrastructure growth on the development of the lower-waged sector and how the effect of it has influenced people's quality of life. We also look at the lower-waged constituent of the mass markets and discuss the importance for growth in telecoms in developing economies. A number of important topics are discussed which include a brief review of poverty and income statistics for different regions of the world and a look at telecommunications penetration of fixed and mobile telephony and Internet in developing countries. Other dimensions that are addressed in this chapter are the correlation between telecommunications infrastructure access and economic growth. We also show the importance of access to basic communications for the mass market and developing countries and show how the communications needs of end users in mass markets are evolving. We then evaluate the cost of communications services and attempt to determine if it is higher in developing countries. An important theme throughout this book is the digital divide between developed and developing countries and how communications can bridge this divide; therefore the term 'digital divide' is briefly defined and explained. Finally, we round off this chapter by presenting evidence on the link between mobile communications and GDP and go on to present some pragmatic examples of mobile applications in emerging markets. The structure for the rest of this book is discussed in the following sections.

Chapter 2: Mobile Telephony – A Great Success Story? Can Mobile Growth be Sustained?

Chapter 2 looks at comparing the reach of fixed and mobile services in various developing markets. It also looks briefly at the prepaid revolution and discusses the unexpected development of the mass market. Further reflections are made on some key issues such as the tendency for operators to focus on high value end users and their lack of understanding of who these end users are. The important role of regulators is also touched on and the inability of regulators to bring down the cost of mobile telephony and create affordable communications for all is critically analysed. The failure of telecoms players to move beyond voice and SMS and their lack of understanding of how to satisfy end users are debated.

We also look at some of the main reasons why customers show little appreciation for innovation and seek to answer why it is that that the majority of customers would be happy to give up the use of advanced services in return for lower call costs. The evolving telecommunication value chain is then discussed and we look at the limits of success for telecommunications development and try to answer the question where will telecommunications go next? We also attempt to provide answers to the question of how growth can be sustained in the developing countries. Operator profitability in the mass market is then discussed in more detail and we attempt to find answers to some important operators' questions such as: can current operational set-ups be sustained by serving the mass market? We also ask: is low average revenue per user equivalent to low profitability? We then go on to discuss the importance of funding and look at the 'Gramen Banking Model' and try to determine if this model can be applied to telecoms using practical examples from Uganda. Finally, we end this chapter by taking a look at operator consolidation in emerging markets and debate whether it is necessary and/or inevitable.

Chapter 3: Communications for All – Is it a Myth?

In chapter 3, the authors determine who the key stakeholders in telecommunications are. The role of government and regulators is evaluated and we try to find out if they represent the interests of the mass market. We then go on to find out if donor communities are really assisting to spread the growth of mobile telecommunications and establish if the poor are subsidising the rich. Thereafter, the role of operators is evaluated to establish if they are greedy or merely victims of their own success, and we argue that excessive profits from mobile telecommunications have trapped operators because shareholders expect them to maintain the same levels of profitability, using case studies from India and the Philippines. The role of mobile service providers is then questioned and we ask and answer the important question: are service providers acting responsibly? Finally, we conclude chapter 3 by looking at the responsibility of the different stakeholders to the cause, and examine the part played by the international donor community in creating affordable communications for all and provide answers to the question: what role should they be playing?

Chapter 4: Customers' Needs for Telecoms Services and Applications

In chapter 4, we evaluate whether consumers in developing countries use the phone in the same way as those in developed countries. Relying on evidence from Ericsson's Consumer Lab, the services that people in developing countries demand are presented. We also look briefly at the type of applications that may be exclusively relevant for developing markets. The provision of service is not without challenges and we take a look at some of the challenges associated with penetrating a market with advanced technology and services including access to handsets, maintaining charged batteries, the fear of technology among consumers and creating and maintaining a presence of local content providers.

Chapter 5: Mobilising Wireless Communications for Mass Markets

In this chapter, the author looks at the business case for investing in rural communications. The role of the donor community in bringing affordable communications to the mass market is explored. Then, the different technology innovation concepts including micro base-stations, shared sites, refurbished equipment (reuse of old GSM equipment), going straight for 3G (more efficient), low-cost handsets, service innovations, airtime transfer, increasing customer loyalty, simplified pricing, loyalty programs and how these can be applied to the mass market are discussed. Also discussed in this chapter are some community innovations such as the rural village phone concept and the commercialisation of rural mobile communications. Finally, this chapter takes a look at the important role that Universal Service Area Licensees (USALs) have on rural telephony penetration.

Chapter 6: Defining Innovative Business Models for Sustainable Telecoms Growth

The liberalisation of communications has resulted in the entry of new communications operators and service providers. As a result, competition has become aggressive and has intensified in the mobile communications world. Operators are looking for new ways to remain profitable and create shareholder value. In this chapter we examine how competition is changing in this sector and look at how the monolithic operator are changing towards being lean and mean. A number of questions are raised here such as: Should service provider's focus on their core business? If so, what really is their core business? How will telco's sustain profitability in the future? What services and applications will enhance quality of life for all? Exploring different growth models to reach out to the poor and most deprived. We also look at the role of mobile communications in social upliftment and future economic growth.

Chapter 7: Straight from Top Executives – Trends and Approaches

In this chapter, we take a look at some of the key issues facing telecommunications CEOs and attempt to encourage debate about where they think the mobile industry is going. We also take a closer look at who should be responsible for promoting growth in mass markets and question whether telecoms will follow the same fate as the aviation industry. We then go on to question what role mobile technologies should play in ICT development. Other important questions that are addressed here are: are industries merging? Is convergence convenience? And is there a business case for mass market growth?

Chapter 8: Internet in Rural Areas – Emerging Business Models and Opportunities in Developing Countries

Chapter 8 looks at the Internet in rural areas and highlights some interesting and relevant Internet business models so critical for the provision of information, knowledge and Internet based services to rural communities. Fresh advances in the field of information and communication technologies (ICTs), mainly computers, electronic networks and Internet based applications along with traditional forms of communication media (such as radio, telephone, television, etc.) are the main means for empowering the poor, uplifting skills, enhancing productivity and enabling them to participate fully in the global economy. In this chapter we look at the current status of the Internet in rural areas and highlight the poor Internet penetration levels that exist in rural areas of the developed and developing countries. We go on to provide some examples using real life case studies to show how the Internet can be used to disseminate information on prices and opportunities, improve participatory decision-making processes, governance at all levels and to provide effective and efficient delivery of government services. Throughout this chapter we highlight the significance of Internet based services in providing both immense opportunities and challenges in meeting these increasing demands and in improving the lives of the rural poor. Finally we round off this chapter by focusing on the important role of the Internet in rural areas and present a few emerging business models and opportunities for developing countries.

Chapter 9: Making it Happen – Enabling Communication in Developing Economies

In this last chapter, the important role that telecoms play in global initiatives for development in trade, poverty elimination, aid programmes, etc., is discussed. The importance of using telecommunications as a tool to empower local communities is highlighted. Thereafter, the roles of other technologies like VOIP, MOIP etc., are pointed out and discussed. Finally, we look at what is needed to build the 'all-communicating world.'

Acknowledgement

The process of writing a book on a unique subject is a highly stimulating and collaborative experience that involves the support, dedication, commitment, efforts and responses of many people. This book is the outcome of many years of work on an important theme that goes right to the heart of creating an all communicating world, made possible today through richer means for people in even the remotest corner of the world to communicate. This book sends out a strong message that mobile communications as an instrument of progress will improve the lives of ordinary people. It aims to show that the means exist and that now all that is needed to reach this goal is the commitment and will of all the people involved.

In our endeavour to promote communications for all, our indebtedness extends to the will and minds of the many people who have shaped our thinking and who contributed to our ability to write this book.

At the outset, we take this opportunity to thank Ericsson's executive management team and in particular Hans Westberg for placing their trust in us and being a great source of support and inspiration. We are also very grateful to Dr K. Dappa who has been a sharp trigger and stimulus to the thinking processes that eventually led to the realization of this research project.

Our initial analysis of the hard facts and figures was daunting and highlighted the fact that the poorest of the poor were paying the highest price per minute for mobile calls. These are the same end users who are not afforded the luxury of any subsidies to promote their inclusion in an all communicating world. We needed to plant the seed to show what could be done to draw the poorest of the poor into the wonderful world of communications. Relying on a number of pragmatic best-practice business models, our primary motivations were driven by the need to show how it was possible to bring communications access to all of humanity and to bring to light the fine balance needed to ensure the sustainable growth of the communications industry. In this regard we acknowledge and extend our sincerest gratitude to Professor Das for clarifying a number of intriguing issues in the ever changing communications business model.

A special word of thanks also goes to Professor Johan Wilhelm Strydom, professor in Business Management at the University of South Africa, for critically reading and making suggestions on some of the chapters.

This book would have been incomplete without the thought-provoking and down-to-earth examples of how mobile communication is impacting the lives of the poorest of the poor. Therefore, we would especially like to extend our gratitude to all our colleagues from Ericson ConsumerLab for sharing these examples with us. A special word of thanks goes to Eric Cruse who sacrificed precious time and effort to share his extensive research work in a number of developing countries. We also would like to thank Henrik Pålsson for reviewing the contents of the book and giving his valuable input.

We would like to thank all our colleagues from Ericsson Business Consulting for sharing their vast experiences and knowledge gained from working with the numerous senior executives in the Information Communications Technology (ICT) industry.

Much appreciation is owed to Katarina Mellström who has been a great supporter and championed the initiation of this project.

We also wish to acknowledge the efforts of the external and internal reviewers too numerous to name individually, who read earlier drafts of this book and suggested many insightful improvements. Special thanks are given to all those who provided invaluable help during the final editing, translation, layout and material production of this book.

Authoring a book in one's spare time inevitably takes precious time away from the people who are dearest to the author. Without the unwavering support, encouragement and dedication of these people, this book would never have been completed. Hence we would like to extend our special thanks and love to Anita, Kerstine, Hawa and Jayasri, our dear wives, for their endearing spirit and faith in us, and for the immense sacrifice that they made during the many weekends and late nights that we each spent researching and writing this book.

And finally, we'd each like to acknowledge the value of our collaboration, and its resonance, in the writing of this book.

Disclaimer

The authors have drawn on their international experience in the development of information and communication systems from their work with Ericsson and the World Bank. The book has been sponsored by Ericsson with Sanjay Kaul leading the effort.

The findings, interpretations and conclusions expressed herein are entirely those of the authors and do not necessarily reflect the views of Ericsson or the World Bank. Ericsson cannot guarantee the accuracy of the data included in this work. The boundaries, figures, denominations and other information shown on any pictures in this work do not imply on the part of the Ericsson or the World Bank any judgment of the legal status of any territory or the endorsement or acceptance of such boundaries.

About the Authors

Sanjay Kaul

Sanjay Kaul is a hardcore telecom professional with over 15 years of telecom industry experience. He is currently working as Vice President for Multimedia & Systems Integration at Ericsson Sub Saharan Africa. Prior to this Sanjay held a number of management positions with Ericsson.

Sanjay started his career in 1992 working in the telecom domain; he has a vast amount of experience in telecommunications and has worked on a number of telecom start-up projects mainly in Asia Pacific and Europe that included a number of underdeveloped and developing markets. He has been involved in a number of management planning, strategy and operations development projects for both mobile and fixed operations, in a management consulting capacity. He has been an active contributor to a number of magazines and has written a wide range of publications for prestigious industry forums such as the *ITU*, *Telecom Management Forum*, *Wireless Communications*, *IIR*, *Business Brief*, *CFMRR*, *Connect World* etc.

Sanjay is quoted and has been commenting on a number of hot industry issues and has been a keynote speaker and chair in a number of industry forums and discussions.

Sanjay is a Telecoms Engineer, MBA in Marketing and has recently concluded his Management & Leadership Education at Harvard Business School.

Dr Fuaad Ali

Fuaad Ali is Managing Director at Habibi Enterprises International, South Africa having expertise in Business Planning and Strategy, as well as Marketing and Services Developments.

Fuaad has 20 years of telecommunications experience. Before joining Ericsson, Fuaad worked for Telkom SA, the largest fixed line operator in Africa. During his period of employ with Telkom he held a number of senior management positions in strategic areas of Marketing, Business Development, Government Relations, Human Resources and finally Strategic Planning. In the period he has been with Ericsson South Africa, Fuaad has done consulting work for Orange, Mascom and BTC in Botswana, Sentech South Africa, Vee Mobile Nigeria, MTN and Nitel Nigeria and has international experience of mobile and fixed markets in Africa through his projects with African operators. Fuaad has travelled extensively in Africa and has first-hand experience of the African fixed and mobile communications markets.

Fuaad is a pragmatic person with strong acumen in communications marketing, strategy and human resources development. He has excellent team and people skills. He is able to build solid relationships with customers and project team members alike. He was instrumental in selling and delivering multiple business consulting services to mobile and fixed line operators in Botswana, Nigeria and South Africa.

Fuaad is also the author and co-author of papers published in accredited journals and he has written a number of guides in marketing for the University of South Africa. He is also one of the authors for a textbook in International Marketing that is presently in print with Oxford University Press, South Africa. Fuaad has also chaired a number of organizational and government forums in South Africa.

Ram Janakiram

 Subramanian (Ram) Janakiram, is Champion, ICT for Rural Development of the E-Development Thematic Group in the World Bank and a specialist in rural information and knowledge systems and rural development. He led the design and implementation of the rural information and knowledge system in Russia as part of a World Bank agricultural project – which is considered as a successful example of the World Bank's operations in this area and is being used as a case study in Cornell University, USA. He has over 30 years experience in the design and implementation of rural development projects in over 30 countries in Africa, Asia, Europe and Central Asia, Middle East and Latin America.

He developed the strategy for World Bank cooperation with the Department of International Development (DFID) of the UK Government and the Food and Agricultural Organization (FAO), for the collaborative program to develop a common framework and approaches for information and communication for the development of rural livelihoods (ICD) at policy and operational levels. He represented the World Bank in the joint WB–DFID–FAO missions to Armenia and Vietnam and identified areas for Bank intervention in the ICD program. He has led discussions on information and knowledge management issues in rural development operations and on e-development for economic growth and poverty reduction during the 2004 Program of Seminars in the World Bank–International Monetary Fund annual meetings He has made several invited presentations in national and international conferences on the role of information and communication technologies in rural development and e-government.

He obtained a spot award for an outstanding contribution to the World Bank's Information Solutions work program in 2004. He holds a Master's degree in water resources management and agricultural economics from Colorado State University, Fort Collins, USA and Bachelor's in Mechanical Engineering from Regional Engineering College, Durgapur, India. His work philosophy is teamwork and consensus building to achieve results.

Bengt Wattenström

Bengt Wattenström is Director of Business Development at Ericsson where he has been since 1995. During his time at Ericsson he has worked in various positions and in different parts of the world. The last couple of years he has focused on developing new Business Models especially for emerging markets globally. One example of a new Business Model is being set up in Tanzania.

Before joining Ericsson Bengt worked in the Computer industry for 25 years. Bengt worked for companies such as IBM, Digital Equipment and Computervision. He has during this time learnt a lot about how market changes impact corporations and their ability to develop and survive.

Bengt has an extensive background in building/driving Services business's internationally both in the IT world and the Telecom world. This includes everything from developing Business/Strategies to implementing them. This means working both at central and local levels to drive the business cultural change from a box selling Company to a Solution selling Company. During his career Bengt has spent a majority of his time in Services, developing Services to a Business in it's own right.

Bengt has an educational background from Harvard Business School as well as INSEAD in France.

1

The Impact of Communications on Developing Markets

OVERVIEW

In this chapter we review the impact of telecoms on developing markets. We briefly look at the importance of and need for technological progress in developing countries and a number of important topics are discussed here including the importance of information access for developing countries, developing countries and Internet access and the need for new business models to increase communications penetration and access in developing countries. The correlation between telecommunications and economic growth is also discussed and we examine how the communications needs of end users in mass markets are evolving. An important theme throughout this book is the *digital divide* between developed and developing countries and how mobile communications can bridge this divide, therefore the term digital divide is defined and explained. Then, we look at wireless communications in developing countries and finally, we round off this chapter by looking at the role of business, government, non government organisations (NGOs) and other donor organisations and try to determine if their efforts are focused.

The world consists of vast continents endowed with enormous human resources. Despite these endowments there are still many countries whose information and communications infrastructures are broken down and where communications development lags behind the developed world. Today, however, the majority of these developing countries are awakening to the innovations in technology and are rapidly taking up the challenges

Business Models for Sustainable Telecoms Growth in Developing Economies
S. Kaul, F. Ali, S. Janakiram and B. Wattenström
© 2008 S. Kaul, F. Ali, S. Janakiram and B. Wattenström

confronting them by undertaking development towards economic and information communications reform.

In the past two decades, developing countries have come to realise that if they fail to catch up with the quick pace of information communications technology development that is taking place in the developed world, then they will be left out of the mainstream global economy and may face even further hardships in the future. As a result, most developing countries realise the need for rebuilding their transport and communications infra-structures. In recent times, the operating and business environments in developing countries have witnessed a number of significant transforma-tions. One such transformation is the *opening up* or liberalisation of markets that were previously dominated by monolithic incumbents. At the same time, a number of regulatory bodies have been formed to help sustain and facilitate market competition in developing countries. Despite these changes, there is still a huge deficiency in the number of people that have access to communications in developing countries, mainly because there is strong misperception amongst communications operators and service providers that little value can be extracted from mass markets in develop-ing countries.

DEFINING DEVELOPING COUNTRIES

One of the main difficulties in assessing the impact of communications on developing countries is that such countries are very different to one another. Developing countries differ widely in terms of the degree to which they have introduced Information Communication Technology (ICT) and the extent to which they have the necessary capability to exploit communications technology. As a result, there is a strong likelihood that there is a group of countries for which some statements made may be either not meaningful or not applicable.

Nonetheless, there are a number of different criteria that are available for deciding whether a country can be considered as a developing country or not. These definitions are generally tied in with a country's right to receive development aid under the rules of a multilateral or bilateral agency. Countries have been classified differently by agencies. Some agen-cies may classify certain countries as developing while others may classify these same countries as developed. Sodowsky (2006) strongly argues that all countries are partitioned into developed and developing sections, meaning that there are developing components in all countries including the so-called developed countries. Citing the United States of America as an example, he goes on to say that there are developmental imbalances between the regional urban and rural areas of the USA which would cause one to identify some areas as '*developed, developing, or, worse, underdeveloped and not necessarily developing'*. It is clear then that unequal developmental distributions exist in all countries.[1]

Certainly, within this framework, extending communications penetration and access to developing regions of the world means that these activities are taking place in all developing areas of countries across the globe. The classification of a country, however, may be less relevant than the degree to which the lives of those who do not have access could be improved by having it.

IMPORTANCE OF INFORMATION ACCESS FOR DEVELOPING COUNTRIES

Some of the major problems confronting developing countries are widespread poverty, poor literacy levels, unaffordable foreign debt, overpopulation and a strong reliance on primary economic activities.

The 21st Century is characterised by new economic dynamics. The currency of wealth creation in this new era is information. To succeed in today's competitive world, nations and individuals alike need access to information. Information plays a crucial role in the alleviation of these problems. Information, mainly in the agricultural, commercial, medical and technological areas can boost the commercial, social, economic and technical infrastructures needed to support the development process.[2] The possession and use of knowledge is vital for making progress. Hence, information access is a key requirement for economic and social success in today's world where borders are fragmented. Information access is indispensable in the logical use of resources, scientific and technological advancement, agricultural progress, industry and services development. *'Consequently, assimilation of scientific and technological information is an essential precondition for progress in developing countries.'*[3] A formidable problem for developing countries is the availability, access and distribution of information to the general population, the so-called *mass market*.

It is clear that the provision, distribution, access and easy availability of information is critical to the success of any nation state. The main role that communications operators and service providers play in society is facilitating the exchange of information (even in its most basic form as

[1] Sodowsky, G. 2006. http://www.isoc.org/oti/articles/1196/sadowsky.html
[2] Heitzman, J. 1990. Information Systems and Development in the Third World. Information Processing and Management 26(4) (), 489–502.
[3] UNESCO, *Draft Medium-term Plan (1984–1989)*. Second part, VII. Information Systems and Access to Knowledge. General Conference Fourth Extraordinary Session, Paris, 1982, 157.

voice) between end users. This vital role of providing access to information is therefore directly affected by globalisation that has resulted in the rapid spread of technology and ideas at speeds the world has never witnessed before. Telecommunications is the central link between the points where information (knowledge) resides and where it is accessed. The Internet, which is a global knowledge repository, has enabled the world to be connected twenty-four hours a day, seven days a week.[4] The easy setup and distribution of mobile communications makes it the ideal tool to enable developing countries to enter into the 21st Century information society. A major restraint, however, is the prohibitive costs of communications access for mass markets.

DEVELOPING COUNTRIES AND THE INTERNET

All countries face substantial challenges. Among these is the need for them to provide their people with good health systems, relevant education, opportunities for personal development and growth, proper housing, employment and sufficient income to meet material needs. Although individual countries may disagree on how to go about achieving these goals, there is widespread agreement about what these goals should be.[5]

Rapid expansion of the Internet holds enormous promise for developing nations. Developing countries stand to benefit greatly from the Internet's communication and information delivery capabilities to help meet their growing needs. The rapid transformation of hard base information to electronic media is also making global information resources easily available to a global audience through the Internet. Developing countries can benefit from this information revolution through communication and information access. However, in sharp contrast to the situation in the developed world where transport and communications infrastructures for delivery of both physical goods and information services are well established, the choices available within developing countries are either generally slow, expensive, or totally nonexistent.[6]

The correlation between information, communication and economic growth is well-known, lending strong support to the usefulness of information networks. Electronic networking is a powerful, rapid and inexpensive way for people in developing countries to communicate, exchange and grow information value. In those areas where communications networks are readily available, communication and information collaboration takes place naturally. Today, the fast pace of decision making required

[4] Gerrard, A.L. 2006. http://www.slis.ualberta.ca/issues/alg1/global.htm
[5] Sodowsky, G. 2006. http://www.isoc.org/oti/articles/1196/sadowsky.html
[6] Sodowsky, G. 2006. http://www.isoc.org/oti/articles/1196/sadowsky.html

in the global environment increases the need for networking and collaboration to enable scientific research and other development efforts which generally lead to tangible economic benefits. Commercial economic growth feeds off widespread communications networks. It is enhanced by access to information and improved contact with support and purchasing personnel, suppliers as well as customers. Access to electronic information networks plays a decisive role in contributing to the improvement and overall effectiveness of the developing communities, comprising representatives of international agencies, non-governmental organisation staff and others working locally and abroad. In addition, many universities and other educational institutions in developing countries are focusing on curricula that might contribute more directly to economic growth and network connections for administrators, professors and students are already important.

Finally, as has been demonstrated in a number of countries recently, there is a strong link between the free flow of information and movement toward democratisation. This cannot be downplayed. Access to information affects political democratisation efforts at a global, regional and national level and influences democratic thought within nation states. In developing countries, where much of the media is often controlled by the state and individual access to networks is currently limited, the need to decentralise control over information and over networks themselves is clear.

PROBLEMS DEVELOPING COUNTRIES FACE

In general, within developing countries, specialised knowledge is often either missing or in short supply. There is generally substantial competition for the fewer more qualified individuals within both the public and the private sectors, as well as between them. Emigration to better labor markets (the so-called *brain drain*) causes erosion of the resources necessary to exploit technology, in the face of countries having a limited set of human resources with which to work.

In developing countries, information poverty is one of the more significant and subtle obstacles to effective exploitation of information processing and other types of technology. Lack of adequate information regarding developments in other countries and other environments is often unnoticed, and in the absence of new information and ideas, old ways of working continue without awareness of better alternatives. Furthermore, developing countries are directly affected by a lack of information hence, social progress and growth are seriously impeded.

Most developing countries are financially poor, relative to developed countries. They suffer from low levels of both financial assets and national

income. Their economies are subject to wide-ranging performance fluctuations due to factors beyond their immediate control. Some are not
viable without sustained development aid.

Many developing countries are benefiting from direct assistance in
obtaining computing and networking technology. Involvement with
private sector firms in developed countries can have significant benefits;
policies promoting domestic investment as well as taxation and profit
repatriation incentives can encourage firms to enter local markets and
provide benefits for a country. Private foreign investment in high-technol-
ogy fields often brings significant flows of information and competence
development opportunities.

Other contextual constraints may apply. Information and financial
poverty and strong governmental misperceptions about the costs and
benefits of network connectivity very often result in long delays to investment in and promotion of communications development activities. In fact,
for these reasons communications spending may be considered very
expensive in relation to the countries' other needs. Another major stumbling block to information and communications development is the lack
of shared vision and synergy between the developing countries' multiple
stakeholders.

COMMUNICATIONS AND THE VALUE OF INFORMATION

Those countries that do not have access to advanced communications
media fail to benefit from the richness of information, knowledge and
value that is created by advancements in telecommunications technologies. As shown in figure 1.1, there have been a number of stepped advancements in mobile communications technology. First generation (1G)
networks, due to slow processing speed, are only capable of carrying
voice across the network platform, thus limiting the richness and value
of communication content. Second generation (2/2.5G) networks have
the added capability to transport basic data across the network, thereby
enhancing communications content. Third generation (3G) technology
was developed to accommodate the convergence of voice, data and image
(multimedia) and in its converged form represents the highest order of
communications content. Clearly, the value of information and knowledge transmission increases as mobile technology advances.

Figure 1.1 shows the relationship between mobile communications and
information and knowledge growth. As can be observed from figure 1.1,
there are different levels of value for information and knowledge and
more advanced mobile technology such as 3G (Third Generation) and

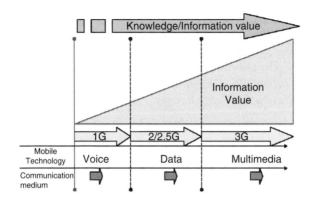

Figure 1.1 21st Century Information Society Value

HSDPA (High Speed Digital Packet Access) enables the highest level of information and knowledge to be commuted across the mobile network enabling the value of information and knowledge to grow from basic voice to data and finally multimedia applications.

In the developed world, fixed and mobile communications technology has reached a mature stage of development. All developed countries have multimedia technology capability. This has resulted in widespread delivery of information and strong growth of knowledge in these countries and an overall advancement of society in all spheres.

TECHNOLOGICAL PROGRESS

In most developing countries, information communications technologies are outdated. Fixed line penetration is very low and it is becoming increasingly clear that if developing countries are to catch up in the communications race and take up their place on the global platform, they will have to do so by relying on wireless technologies. This suggests that new communications business models should ensure that wireless technologies are intertwined as part of the developing region's technology of choice and that they are affordable.

Another critical challenge confronting developing countries is the shortage of network capacity. The slow but increasing subscriber numbers requires increased network capacity, thus opening new opportunities for telecommunications operators and service providers. Presently, the mobile infrastructure that exists within the majority of developing countries is constrained by a lack of new investment and the high numbers of users sharing limited available capacity. This has resulted in network overload and congestion during peak hours and in some places such as Botswana and Nigeria, a serious reduction in quality of service. Furthermore, the existing technology available is limited to provide voice and in some cases basic data services. This does not bode well for developing countries since it limits their access to quality information and knowledge creation and widens the knowledge gap between the developed and developing world.

WHY NEW BUSINESS MODELS?

One thing is clear, if developing countries are to enter into the information age they will need innovative and disruptive business models that promote and stimulate communications adoption, penetration and promotion. These models must not only provide communications solutions for the rich, but must also provide for the poor in mass markets.

It must also present a good business case for investors that promotes a viable business proposition and makes good business sense for them to undertake communications investment in developing economies. These business cases must ensure that mass communications penetration is achieved, thereby boosting subscriber numbers in mass markets and ultimately, long term sustainability and profitability inherent in economies of scale. Business models that are provided should promote affordability and adoption by the network service providers in these countries and ensure that they will fit the needs of a much wider range of communications end users. It is widely accepted that the majority of communications end users in developing economies are not high revenue generating users of communications and have a low ARPS (Average Revenue per Subscriber). This by itself implies that innovative business solutions are needed to make up for the lower levels of revenues to compensate profitable operator investment in communications. It is interesting to note recent evidence by the ITU suggesting that mobile average revenue per user in some developing economies of Africa is rising.[7] This would certainly seem to imply then that there is a strong business case to provide communications to mass markets.

It is evident that any increase in mobile subscriber numbers will increase the need for greater network capacity. The existing network capacity of developing countries is already overburdened and the shortage of network capacity will curb future growth prospects. New infrastructure and innovative financing solutions are therefore needed to ensure that demand can be met. The demand for telecommunication services is clearly there, and

[7]ITU. 2004. http://www.itu.int/AFRICA2004/media/wsis.html last accessed on June 23, 2007.

to take advantage of it requires innovative, simple and specialised business models that match the needs of the different developing economy markets in relation to costs, technology, services, infrastructure and marketing.

TELECOMMUNICATIONS PENETRATION AND ECONOMIC GROWTH

Given the wide range of other problems afflicting developing countries (for example, poverty, AIDS and starvation) these countries are faced with various economic choices. Any form of investment has to be based on achieving the highest return for investment. This is an important factor that requires serious attention. If developing countries are to divert their financial resources towards communications development, they will require a strong business case supported by empirical evidence that demonstrates the benefits attached to communications infrastructure development. Such a business case must clearly demonstrate that the benefits associated with communications infrastructures spend surpass the opportunity cost of the other choices that are open to them. Alleman et al.[8], argue that there is a high correlation between telecommunication penetration and economic growth.

Many countries have introduced far-reaching reforms into their telecommunications sectors. In developed countries, establishing a highly competitive communications market has been at the epicenter of regulatory reform. In the developing countries on the other hand, regulatory reform objectives have been aimed at the overall improvement and enhancement of making communications services accessible across a much wider distribution of the population. Additionally, across all countries, there has been a greater reliance on private sector capital rather than public capital for the development of the telecommunications, and there is a dominant need for greater flexibility in order to adapt to rapid technological advances such as in wireless telephony, the Internet and Voice over Internet Protocol (VoIP) telephony.

A modern well-established telecommunications infrastructure sets out the basic foundations for economic growth for both developed and developing countries.[9] The critical role of infrastructure capital (for example, highways, water and sewer lines, etc.) in the economic development of a country has long been recognised. Researchers who have found that communications investment directly impacts economic growth are only

[8] Alleman, J. 1994. Telecommunications and economic development. Evidence of causality. Case study.
[9] Chan, N. 2006. Wireless Mesh Networking for Developing Countries.

recently supporting this position. In a landmark study, Roller and Waverman[10] showed that telecommunications investment in infrastructure both directly and indirectly impacts on the economic growth of countries. However, these authors are quick to point out that the effects of telecommunications on economic growth are only observable after a country achieves a certain level of critical mass.[11] Interestingly, the critical mass needed to influence economic growth, according to Roller and Waverman, only becomes noticeable when teledensity reaches 40 main telephone lines per 100 population. This means that the effects of telecommunications on economic growth are only felt when telecommunication penetration reaches a level associated with universal service.[12] It also means that if developing countries are to realise economic benefits from telecommunication penetration they must accelerate efforts to increase penetration and mobile infrastructure is the best way of doing this in a relatively short time period.

A number of researchers[13] argue that there is a strong positive correlation between telecommunications and economic growth in developing countries. Generally, growth in telecommunications contributes to economic growth mainly because it generates spin-offs due the multiplier effect. In Nigeria, South Africa, Kenya, Uganda, India, China and Indonesia for example, growth in telecommunications has led to the creation of a vibrant secondary market. When traveling in Nigeria, especially in the main cities of Lagos, Abuja and Port Harcourt, it is not uncommon to see hundreds of umbrella stands distributing mobile airtime. At the same time, telecommunications growth has enhanced employment opportunities for many who would otherwise be unemployed. It has also contributed to the growth of local industries supporting the mobile telecommunications infrastructure providers. In the rural areas of developing countries, the exchange of goods and services has been facilitated by the availability and access of mobile communications, thereby contributing to and enhancing the promotion of economic developments in these areas. This is increasingly helping to improve the overall quality of life amongst poorer rural communities.

[10] Roller, Lars-Hendrik and Waverman, Leonard 2001 Telecommunications Infrastructure and Economic Development: A Simultaneous Approach, *American Economic Review*, September, pp. 909–923.
[11] Onwumechili, C. 2001. Dream or Reality: Providing Universal Access to Basic Telecommunications In Nigeria? *Telecommunications Policy* 25(4), 219–231.
[12] Roller, Lars-Hendrik and Waverman. Leonard 2001. Telecommunications Infrastructure and Economic Development: A Simultaneous Approach, *American Economic Review* September, pp. 909–923.
[13] Norton. S. 1992. Transaction Costs, Telecommunications, and the Microeconomics of Macroeconomic Growth, *Economic Development and Cultural Change*, pp. 175–196.

Economic development and investment in telecommunications infrastructure go hand in hand, as evidenced by many developing countries in Asia, Africa and Latin America. To some extent, the network effects resulting from increased investment in telecommunications infrastructure may have accelerated economic growth in these countries. Another benefit of infrastructure growth in developing countries is the possibility of reduced international call imbalances leading to more balanced international rate settlements between developed and developing countries.[14]

COMMUNICATIONS FOR MASS MARKETS IN DEVELOPING ECONOMIES

The growth of telecommunications access is a major step towards global connectivity. This means the development for rural communities in developing countries where about 70% of the population resides. Historically telecommunications ownership has been limited to the urban elite, and today a high priority for developing countries is to ensure that telecommunications infrastructure development is targeted at the mass market that is not concentrated in the urban centres. Among those developing countries at the forefront of providing telecommunication access to remote areas are India, Iran, South Africa, China, Tanzania, Zambia, Botswana, Mozambique, Ghana, Nigeria, Malawi, Kenya, Ghana, Senegal and others, each of which is directing significant planning and more funding to this effort. On a much larger scale, global connectivity implies access to sophisticated international networks, specifically the Internet. In addition, access to international networks also means access to information and knowledge that generally lead to an improvement in quality of life. Unfortunately, there is still a huge *digital divide* that exists in the world today.

THE DIGITAL DIVIDE - WHAT IS IT REALLY?

Recent research shows that providing access to Information and Communication Technologies (ICT) drives economic growth, improves business, education and employment opportunities and facilitates social development by allowing people to keep in touch with family and friends. Many people are denied this access, mainly in developing countries. The gap that exist between those that have and those that do not have access

[14] Hardy, A. 1980. The Role of the Telephone in Economic Development, *Telecommunications Policy*, 4(4), 278–286.

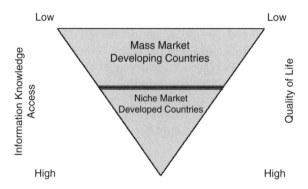

Figure 1.2 Knowledge and digital divide impact

to information, knowledge and enabling ICT technologies is generally referred to as the *digital divide*.[15] Closer scrutiny around the reasons behind these widespread disparities shows that higher level middle class societies have access to high quality digital technology and that the reason for this is primarily because those behind the drive are motivated by profit. Technology developers develop customised technology solutions for the upper middle classes because they believe that money is to be made from this class. Unfortunately however, the last 2 billion, the poorest of the poor, have been excluded because there is a mistaken belief by technology developers, that any solution developed for the poor is unprofitable. In addition, in those places where the poor have been given access to ICT, the quality of the service is generally inadequate.[16]

As depicted in figure 1.2, mass markets in developing countries experience low access to information and knowledge directly as a result of the digital divide. Furthermore, people in developing countries experience a lower quality of life, compared to those of the developed rich countries.

Human life has equal value. If mankind is to improve the quality of life for all and if new markets for growth are to be found, development of ICT for the poor must take on a new approach that involves a sharing of responsibility between all the different stakeholders (see chapter 5). By providing the poorest of the poor with quality technology solutions, the poor will benefit from advanced digital technology that is only benefiting the rich.

[15] Canning, D. *Telecommunications Infrastructure, Human Capital, and Economic Growth.*
[16] Duwadi, K. 2003. *Telecommunications Investment, Economic Growth, and Universal Service in a Global Economy*, International Bureau, FCC.

CLOSING THE DIGITAL DIVIDE – WHY IT MATTERS

There are various reasons why closing the digital divide really matters. For one, it is evident that closing the digital divide is a necessary precondition to reduce poverty worldwide. There is a strong consensus among some antipoverty proponents who believe that closing the digital divide is not the most important priority, because they feel that the main immediate needs of the poor is to obtain clean water and find work, and that access to information and knowledge is a less pressing need. This view fails to recognise that having access to digital technology promotes the economic and effective efforts directed at improving the water supply, rural health, transport infrastructure, education, and stimulates job growth while at the same time addresses the multiple problems related to poverty. Undoubtedly, closing the digital divide gap may not directly alleviate poverty in the short term, but without closing the gap there is certainly less chance that global poverty will ever be reduced.

The removal of the barriers that support the digital divide is a precondition for achieving sustainable world markets. During the late 1990s when the dot.com bubble burst, many people believed that the digital information industries were a failure and would never occupy a pivotal role in the global economy. When we think of the future today, we know that their belief was wrong. Retrospectively, it's easy to observe that the technology bubble burst merely highlighted the saturation that was taking place for technology purchases in developed markets. Clearly, technology purchase dynamics have changed and the developed world today, North America and Europe are no longer the dominant technology consumers they once were. In fact, it is the emerging countries such as China, India and other Eastern and African countries that are driving the digital economy assisted by the rapid growth of wireless networks. This is also spearheading economic growth in a number of other areas beyond the major urban parts of these countries. The penetration of widespread broadband networks into rural areas in developing countries will lead to reduced costs through economies of scale. Communications and IT companies will do well to review their thinking on closing the digital divide. Through encouraging and promoting new communications and information technology innovations for developing country development, they could help fuel economic developments in developing countries and promote new growth in the developed countries. One very promising area is wireless communications.[17]

---

[17] See www.itu.int/ITU-D/digitaldivide last accessed on April 23, 2007.

WIRELESS COMMUNICATIONS IN DEVELOPING COUNTRIES

Wireless communications networks are enabling millions of poor people around the globe with no access to postal services or fixed telephone services to become connected to the wider world. Wireless mobile phones account for about three quarters or more of all telephones in the developing world and are extremely valuable to individuals in developing countries because other communication forms such as railroad, postal systems, roads and fixed-line phones are inadequate. Wireless networks enable people to plug into the global communications network and integrate into global society by providing them with a unique connection point. Through this, users are able to tap into the global communications network and participate freely in the global information economy. In some cases, people who cannot afford to personally own wireless mobile devices themselves are able to access wireless services through informal sharing of mobile devices with family and friends or by using local community phone shops made possible through wireless networks. Figure 1.3 shows the main categories of wireless applications in developing countries.

Wireless applications for developing countries can be broadly categorised as follows:

- Information access and distribution
- Commercial use

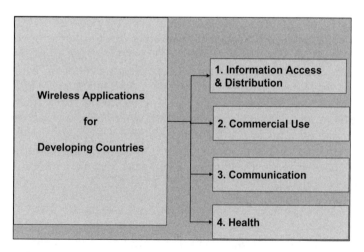

Figure 1.3 Main wireless use categories in developing countries

- Communication
- Health

Each of these is briefly discussed.

Information Access and Distribution

Information access and distribution is one of the most important user application dimensions for poor people in developing countries. Large proportions of people in developing countries are deprived of information access and at the same time, those with information that can improve quality of life are unable to distribute it to them. This is mainly the case because technology developers and providers feel that no value can be derived from this section of the market.

Today, more and more companies are coming to realise that growth in the developed markets has reached saturation and that new value can only be created in developing markets. India, China and Africa are good examples of where the demand for new communications technology is driving growth. Within developing countries, people have realised the importance of having access to information as an enabler to allow them to improve their circumstances. This has led to the mass markets in developing countries demanding access to communications and is pressuring Governments to supply communications access services. Businesses are also recognising the growing importance of the bottom of the pyramid mass market. In Pondicherry, India, for example, a number of information shops using local Tamil language are cultivating interactivity and synergies between the multiple information systems that are available and their intended beneficiaries. Using local content that is developed in conjunction with local communities, information shops are satisfying the information needs of local communities. There are about a hundred databases that include a rural yellow pages which is updated regularly, as the need arises. Using a single website, the entire range of government programmes can be viewed. This enables government in this region to promote more transparency. Appropriate useful information for the local community is assembled on the website. For example, information from the Tamil Nadu Agricultural University, US Navy website and Aravind Eye Hospital can be found on the government website.[18]

In Madhya Pradesh, the Gyandoot project provides another very good example. Gyandoot is an intranet site in the district of Dhar that connects many rural cyber cafés satisfying the daily information needs of the rural

[18] See www.itu.int/ITU-D/digitaldivide last accessed on April 23, 2007.

population.[19] Gyandoot provides the local population with a number of valuable information services such as:

- Commodity information services
- Rural matrimonial services
- Income certificate
- Domicile certificate
- Caste certificate
- Passbook and landowners land rights and loans
- Multiple forms for different government schemes
- Advisory services
- Family listing for below the poverty line
- Employment information
- Rural market information
- Rural news

Case Study 1: *Botswana – land of the emerging new generation African enterpreneur*

Botswana is situated to the west of South Africa which lies on the southern tip of the African continent. The county has a population of about 2 million. In a study conducted by Ericsson during 2005, it was found that two new market segments had emerged: the young enterpreneurs (aged between 16 and 21) and the professional enterpreneurs (aged between 24 and 48). Both of these groups relied heavily on mobile communications to conduct business. These enetrepeneurs used Short Messaging (SMS) to reach their potential customers and to communicate with other business associates. This has resulted in a vibrant enterpreneural environment in Botswana and and increase in jobs in the informal sector. Many of the entrepreneurs surveyed used mobile phones as their only means of communication. All the entrepreneurs in the sample said their business and profits had increased as a result of mobile phones, despite the increased call costs. The Ericsson study also found that mobile phones are used as a community amenity. Most of the mobile phone owners surveyed in Botswana allowed family members to use their phones free of charge. The survey also found that 91% of those surveyed in Gaborone said they had had experienced better relationships and more contact with family, friends, customers and business associates, mainly because of their mobile phones.[20]

---

[19] See www.itu.int/ITU-D/digitaldivide last accessed on April 23, 2007.
[20] See www.itu.int/ITU-D/digitaldivide last accessed on April 23, 2007.

Another area where wireless communications is having a strong positive impact is in the area of promoting commerce and entrepreneurship.

Promoting Commerce

Communications coupled with the intelligence of computers and mobile phones has a strong influence over the commercial strategies of Small Medium Micro Enterprises (SMMEs) and rural entrepreneurs. The impact of this influence is reflected in the areas of natural capital (providing key opportunities for accessing government procurement policies and contracts, financial capital (sourcing information on microcredit organisations), human capital development (increasing knowledge and skills), social capital formation (fostering and promoting the establishment of commercial contacts way beyond the immediate community circle) and physical capital (creating a platform for rural communities to lobby government to provide basic infrastructure. Mobile communications can have a positive and major impact on economic growth. In fact, the economic effects from mobile communications may be twice as large in developing countries as in developed countries. Generally, a developing country with a teledensity of 10 more phones per 100 people between 1996 and 2003 would have had GDP growth 0.59% higher than an otherwise identical country.[21] Generally, small businesses in remote rural areas face serious difficulties in conducting business. They have to travel large distances to access markets or other places where they can distribute their goods. Quite often, they are unable to facilitate advanced arrangements with either buyers or sellers. Mobile phones play a significant role in altering the logistical landscape that rural traders face by enabling mobile-based ordering and delivery requests providing them with the ability to schedule reliable and advance appointment arrangements with business customers and partners.[22] Manobi, the French private telecommunications company in Senegal, West Africa, makes use of Wireless Application Protocol (WAP)-enabled mobile phones to enable Senegalese farmers to access up-to-date real time market prices. This has enabled Senegalese farmers to benefit from the transparency of prices inside the market that many producers lack.

[21] See http://gyandoot.nic.in/ last accessed on April 20, 2007.
[22] Ali, F. 2005. Botswana Mobile Market. A market study conducted for Orange Botswana by Ericsson South Africa.

Case Study 2: *India Shop – using the Internet to reach markets*

India Shop is an e-commerce Internet-based virtual shopping mall that sells Indian handicrafts. India Shop is the brain child of the Foundation of Occupational Development (FOOD) in Chennai. India Shop promotes its local goods over the Internet and through chat rooms and mailing lists. E-marketers work using a computer at home or at cyber café's and draw commissions on the sales that they achieve. The e-marketers respond to sales enquiries and liase with the craftspeople, mainly sharing information using multiple e-mails received from customers before sales are finalised. These e-marketers earn between Rs2000–Rs10000 per month.[23]

Case Study 3: *Accatel and Bangladesh using wireless platforms to reach out with new services*

Chittagong, commercial capital of Bangladesh, a port city with a population of some 3.5 million people, is the place where Accatel and Bangladeshi Nextel Telecom have installed a new IP based wireless mesh network that provides both phone and Internet services to residential and business customers. Accatel was motivated by the fact that in a developing country such as Bangladesh, a reliable wireless mesh network is a fast, reliable and cost-effective way to facilitate and bring more people into the global economy. This network enables the company to profitably provide a bouquet of voice, data, and video services to customers in under-serviced areas where there is no existing phone service infrastructure, without the need to invest in costly traditional communications infrastructure. Apart from being fully wireless, the network is also fully independent of the national electric grid and uses batteries that provide between 4–6 days of electrical power before being recharged through solar panels.

Communications

Communications is the one category that has the greatest impact on society. Human beings are social creatures that are dependent on communications. Communications enables people to share their thoughts, ideas and intentions with others. Every community has a need to communicate both within and externally with others. Telecommunications enables society to bridge the gap over distance, time and space. It is the interactive nature of communications that allows people to share informa-

[23] Vodafone Report, 2006.

tion, grow and develop knowledge and ultimately improve the quality of their lives. All communities have a burning need to communicate. This need can easily be translated to an opportunity that can used to create new value.

Case Study 4: *Extending telecom services to South Africa's poor*

Faced with the immense challenge of providing shelter and such basic services such as clinics, schools and sanitation to millions of South Africans, the added challenge of laying on conventional telephone services to these citizens is a heavy demand for the State. Thus, when Vodacom was granted its cellular telecommunications licence in the early 1990s, one of the conditions was that the company had to deploy 22 000 subsidised public cellular telephones in under-serviced and rural areas which had limited or no access to fixed-line services. Originally, Vodacom adopted two methods to meet its obligation. The first was to issue phones to faculty and administrators at universities and technical colleges in disadvantaged areas. The second was to set up stationary phone shops or kiosks with multiple lines, all connected to Vodacom's existing infrastructure through a wireless link. While the first strategy (transportables) was easier and faster to roll out, it did not achieve its intended goals. Faculties used the phones for themselves and students were rarely given access.

Vodacom phased out all of its transportables at the end of September 2003 and allocated the lines to its 5000 phone shops. Although the regulators do not allow profit from these services, it is presumed that the services must substantially contribute to the recovery of overheads – which is a significant motivation. Vodacom, 35%-owned by Vodafone, one of the world's largest mobile telecommunications network companies, is South Africa's leading cellular communications company, with a 54% market share at the end of December 2003 and providing world-class GSM (Global System for Mobile Communications) services to 10.2 million customers in South Africa, Tanzania, Lesotho and the Democratic Republic of Congo. About 98% of Vodacom's 4300 employees are computer literate and more than one in ten are currently furthering their studies. The company spends almost 8% of its total salary budget on training and development. Vodacom's philosophy is that business should use some of the profit it generates towards ensuring a better quality of life for the communities in which it operates, which is illustrated by the Vodacom Foundation's efforts to develop the company's extensive social investment programme. The funding strategies of the Vodacom Foundation largely follow the priorities set by the South African government.

Case Study 5: *More than micro-finance: Linking business solutions to 'bank the un-banked'*

Geneva, 18 March 2005 – With the UN's International Year of Micro-Credit underway, attention shifts to the exchange and understanding of practical examples and initiatives. In this article the World Business Council for Sustainable Development profiles the business case for microfinance, as well as showcasing examples of work in practice.

Microfinance is regarded as one of the success stories of development. It has demonstrated the capability of communities to look for solutions that create wealth rather than relieve poverty. However, for microfinance to unleash its full potential, and operate at scale, it needs to operate as a business that can compete for capital and not be dependent on free or low interest funds to make it viable. It needs new technologies and new impetus with demand for microcredit, savings and insurance being created from other businesses. This will create the economic virtuous cycle of development that builds the capital of the poorest communities.

Finance is Essential . . . for Everyone

For thousands of years trade has been a source of development, both in sheer economic terms and through the related exchange of thoughts and ideas. It might be argued that in the current era of growing insecurity and poverty the beneficial exchanges brought about through movement of goods, people and ideas are more important than ever. Flourishing trade requires flourishing finance: exchangeable currencies; savings as security and loans for investments. These notions apply to the 2.8 billion people on the planet who live on less than US$2 per day, just as much as they apply to the richer parts of the population.

Business Solutions to Poverty

Member companies of the World Business Council for Sustainable Development (WBCSD) are leading the way in developing an understanding of the opportunities generated by Sustainable Livelihoods business. That is business which benefits low-income communities, whilst simultaneously providing new avenues of growth for companies.

It involves putting the poorest communities at the heart of business thinking to identify opportunities for them to become sources of supply or of innovation. It involves considering how to include and consider everyone as potential employees and customers of the future in a way that makes good business sense. Solutions that engage the poor in local economic activities are likely to be sustainable over the long term and provide growing income and secure livelihoods. The approach to Sustainable Livelihoods business is strictly businesslike because the rigours of a business approach are most likely to yield benefits that are enduring. Business wants to stay in business and in the process can provide employment, can deliver goods and services including basic needs such as water and energy. Moreover business creates opportunities for entrepreneurs as suppliers and distributors, thus driving the virtuous circle of economic growth and sustainable development.

What Lies Beyond Microfinance?

Microfinance lies at the heart of this, since 99 per cent of business starts small and most stays small. Aspiring customers and local entrepreneurs alike often depend on initial start-up capital to enter the marketplace. As we enter the international year of microcredit, the time has come to think beyond the conventional success stories of microcredit. There has been much success and many lessons learnt from microcredit institutions, which often, using foundation or aid capital, have proven that group risk mitigation is a viable technique and that provision of training in basic business skills has to go hand in hand with distribution of credit.

The Grameen Foundation who pioneer and are experts in microfinance reported that the global magnitude of microfinance lending at the end of 2002 was just US$4 billion to 62 million customers. This must be a small fraction of the potential demand, if we assume that at least 1 billion adults are currently 'un-banked'. A loan of just US$100 to each would correspond to a market of US$100 billion.

Three Key Questions

There are roles for many players to ensure that microfinance can flourish and play its full role in development. Three critical issues need to be addressed:
- What does it take to ensure microfinance is sustainable as a business?
- What is needed to ensure demandpull and investment in microfinance?
- How does this apply to microsavings and microinsurance as well as credit?

Encouraging Signs of Expansion

Accessing private investor capital
A number of players are active in work around this value chain. For instance Deutsche Bank, alongside the United States Agency for

International Development and the British Department for International Development have used their core banking skills to develop a US$50 million dollar microcredit development fund which can lend money to commercial banks and microcredit institutions.

The role of the development agencies is to 'backstop' losses that commercial investors would not take. For the agencies, and indeed the taxpayers who fund the agencies, this is an effective way of leveraging donor money to provide private capital and management advice that would otherwise not be available.

Reducing transaction costs
Further along the value chain, WBCSD members Hewlett Packard (HP) and Vodafone alongside the information systems company Oracle are looking to see how their respective technologies can be applied to delivering microfinance services.

HP has been active in providing access to low-cost computer equipment through their e-inclusion and i-community programmes in India, South Africa and Uganda. Vodafone wants to build on its experience through its venture with Vodacom in South Africa and Safaricom in Kenya to see how prepaid technologies and their core skill in managing large back-office billing systems can be extended to southern and eastern Africa.

Visa International and WAY System recently showcased a new technology that allows for transactions to be executed at point of sale (typically taking place outdoors or in local markets in developing countries) through a pocket size Mobile Transaction Terminal. Combining the features of wireless communication with a Visa debit card that allows for tiny amounts to be deposited enables both merchants and customers to complete a secure transaction at minimum cost.

Enabling affordable access to basic needs
Shell in Sri Lanka has recognised that whilst solar power is a good thing – low in running costs, non-polluting and reliable – the up-front capital cost can prohibit ownership of a simple lighting system that might cost as much as a year's income. They have partnered with a local microfinance institution, SEEDS, and developed an offering to meet demand.

Delivering and helping understanding of financial services
Some companies, such as ABN Amro are working directly to develop microfinance banking arms, whilst others are developing business models, which are likely to stimulate demand. For instance Rabobank has developed financial products to take the risks out of uncertain prices paid to rural farmers in West Africa. This helps bring an appreciation of financial services into the poorest communities and provides

a chance for financial institutions to prove their integrity and capability in serving multiple small-transaction customers.

For an energy company such as BP, a partner in the Baku–Tbilisi–Ceyhan pipeline project, working with FINCA (the microfinance institution in Azerbaijan) is an effective way of ensuring that small businesses can develop along the route of their pipeline. Through such a scheme traditional community investment can begin to metamorphose towards long-term business-focused development.The repayment rate for the BTC-funded part of FINCA's portfolio has to date been 100 per cent amongst 4380 clients.

Providing security, when bills are valued
The work of many utility companies in bringing water and electricity to informal settlements – where residents have tenure by dint of being there, rather than through formal title – is another important step. Often a bill or invoice for provision of services is the first step in proof of tenure, which can lead to the resident being able to borrow money to improve his or her property or start a business using property as security on the loan.

Mobilising savings
A natural extension of the technology and the infrastructure for microfinance is to extend it to providing much lower cost and more acceptable transmission of remittances from the diasporas overseas to their homes, and for microcredit institutions to be allowed and able to act as savings institutions. In this way communities can have a stake in their own future and governments can avoid the pitfalls of overseas debt repayments that escalate with sliding exchange rates.

Conclusions

Recent events in Asia have shown not only the vulnerability of populations who have been unable to invest in durable housing, but how whole communities are destroyed through inadequate infrastructure. Microfinance is no panacea, and the private sector is certainly not the single actor that will prevent poverty and crisis. However, in a world of six billion people, of which nearly half live far below any commonly accepted standard of decent living, there is a growing realisation that the dynamics and motivations of business are the drivers of long-term development and security. If set within a fair regulated framework for commerce, business transactions reward efficiency; business delivers essential services such as water, energy and foodstuffs; business creates jobs and income.

Looked at another way, as the world population is predicted to increase to nine billion by 2050, the poorest three billion going on six billion are the employees, the customers and the entrepreneurs of the future.

Microfinance warrants investment and support from all stakeholders to ensure the technology and the regulatory environment is developed to provide products that are affordable, accessible and available at a scale that will make a difference.

There are many opportunities if we invest in 'banking the un-banked'.

Source: Rennie, C. *WBCSD News.* March 18, 2005. http://www.wbcsd.org/ last accessed 4 February 2008. Reproduced with permission from WBCSD.

Making the Grade

Established in renovated shipping containers, phone shops offer telephone services to the public on a cash-per-call basis. These services are growing at such a rate that by March 2001, there were 2335 operating in the country. Some 52 million calls were made from community service phones during March 2001, resulting in some 78 million minutes of call time. In 2003 Vodacom delivered more than 90 million minutes per month through 23 000 active phone lines at 5000 sites, with an annual revenue of US$129.5 million. Vodacom is paid two-thirds of total revenue while the phone shop owner keeps the other third, so each shop brings in an average of US$38 800 per year in gross revenue. These businesses have started producing greater revenues than 'conventional' pre-planned businesses. This shows that when government sets quantifiable social objectives (e.g. providing X number of underprivileged people with access to telephony), business can bank on its creativity and its capacity to innovate to find the most cost-effective ways to achieve the stated objectives. The call shops appear to have a very high productivity/load factor/occupancy – delivering a service that is wanted and affordable in the poorer communities. Vodacom's commitment to providing subsidised cellular telephones in areas with the greatest need has empowered disadvantaged people through partnerships with local entrepreneurs. Approximately 20 000 jobs have been created as each mobile phone shop spawns five jobs. Potentially more jobs have been created from cluster businesses, as well as unquantifiable spin-off economic benefits, including:

* Informal businesses mushroom nearby;
* Entrepreneurs like it – Vodacom has never had someone return a phone shop on the grounds of not making money. Phone shop entrepreneurs are taught business planning and operating skills and learn to become part of the formal tax-paying part of the economy. To iden-

tify suitable local entrepreneurs to run the phone kiosks, the company looked at the phone use patterns of their existing customers. They realised that those who appeared to be using their phones a great deal were most probably renting them call-by-call to friends and neighbors. This clever interpretation of data allowed Vodacom to identify the natural entrepreneurs in the villages, hire and train them to manage the mobile kiosks. Often, these entrepreneurs employ people to assist them, and in some cases revenue they earn from phone shops enables them to expand or to start other businesses. The total cost of getting a phone shop set up is approximately US$7400. Vodacom pays approximately US$3950 to purchase and modify shipping containers for the phone shops. The individual owners are responsible for purchasing the internal equipment and paying for transportation of the container to its final site. The franchisee's total investment is approximately US$3450. Initially, Vodacom set up a financing arm called Vodafin with Future Bank to help entrepreneurs start up call shops. This was phased out as phone shop owners found they could get credit from other sources.

The lesson for business is that, although micro-finance might need initiating, businesses do not need to get locked into the non-core business of running a bank. The company provides an important accessibility step in South Africa's evolving cellular market. Extending communication services within communities that previously had limited access appears to create growing service demand for other cellular products and service areas. The fact that the phones cannot accept incoming calls is an indicator of a key need that personal mobile communications will address within rural and disadvantaged communities. For example, there are individuals with personal cellular phones who receive and pass on incoming messages for neighbours or friends. In this way, customers start using telephones for outgoing calls in the phone shops, which then creates demand for Vodacom's mobile products and services. Entire communities benefit. A local phone shop also provides invaluable services to the residents of a community, including:

- Allowing families that have been separated following employment opportunities in other areas can keep in touch and manage family funds;
- Helping professionals conduct business more efficiently, i.e. delivery truck drivers can call in to their headquarters to report problems or get delivery instructions;
- Allowing individuals to pay bills and order personal effects;
- Allowing access to a range of social services including summoning a doctor or seeking medical advice, as well as gaining access to services such as electricity, sanitation and water.

Scalability

A significant factor in the success of the programme lies in the existing infrastructure that was built to service Vodacom's traditional cellular customers. These users, in essence, subsidise the services provided for customers. Without the traditional services, there would be no infrastructure and the company simply would not be able to justify the cost of developing its own infrastructure, a situation that works well for Vodacom in South Africa. Vodacom has a long-term plan for building on this success and improving the growth prospects of their overall business while maintaining profitability. In particular, there are opportunities to expand the number of sites in South Africa (demand from potential franchisees currently exceeds Vodacom's ability to expand), to introduce new services (such as data and fax) and to replicate in other countries. An important element of the long-term business rationale for this project is that it will introduce the Vodacom brand to a large base of consumers and at the same time, show them the value of a telephone. As they grow wealthier, many may choose to upgrade to a traditional Vodacom mobile phone.

Source: World Business Council for Sustainable Development.[24]

Improving the Health of the Poor

Poor people especially in underdeveloped and developing countries have little or no access to suitable healthcare. Communications in health care is an area that holds enormous promise to alleviate poverty in developing countries mainly because it based on information resources and knowledge. There are multiple ways and means in which communications coupled to computing can be used to achieve good health outcomes. In many developing countries ICTs are being used to facilitate remote consulting, diagnosing, and the treatment of patients who would otherwise not have access to skilled health professionals. Physicians working in remote rural locations are able to take advantage of professional medical skills and experience. For example, the South African Department of Health (DoH) has a *telemedicine* project that facilitates this service between urban medical centres such as Johannesburg General in Johannesburg and rural hospitals situated in areas such as the Polokwane and Mpumalanga Provinces. At the same time telemedicine is being used to provide medical training to medical students at the different medical universities in South Africa. A number of health Internet sites for medical professionals and patients also provide valuable health information that is very helpful to

[24] WCSD. www.wbcsd.org/web/publications/case/vodafone_full_case_web.pdf last accessed on May 31, 2007.

rural poor communities where healthcare is generally insufficient. Centralised databanks connected to fixed and wireless communications networks allow medical professionals to stay abreast of quickly changing medical knowledge.

As a result of advancements in communication and information technologies, a joint initiative by the World Health Organization (WHO) and six of the world's largest medical publishers has resulted in the provision of important scientific and medical information to about a hundred developing countries that would not be able access such information. This initiative allows these developing countries to access top medical and scientific information using the Internet at either no charge or at reduced rates. Before this initiative, medical and scientific journal subscriptions, were charged at the same prices to medical schools, research centres and other institutions, irrespective of whether or not they could afford it. The annual subscription price per journal costs a couple of hundred dollars per title. Subscriptions for most of the important journal titles cost more than US$1500 per year, making it difficult for most health and scientific research institutions in developing countries to access important much needed scientific information.

Case Study 6: *Sipho Zondi and the miracle of communications*

Sipho Zondi is a 42 year old South African male. Sipho the only family breadwinner supports a family of 8 including a widowed mother that is HIV positive. On 14 December 2006, while visiting some friends and family in Polokwane, Sipho complained of severe chest pains and shortly after that he suffered a serious heart attack. He was immediately rushed to the Polokwane state hospital. Dr Rezaan Ali was the general practitioner that attended to him. Having little cardiac experience, Dr Ali quickly telephoned Johannesburg General Hospital and requested to speak to Dr Moti the Cardiologist on call there. Relying on state-of-the-art telemedicine technology Dr Moti was able to examine Sipho and prescribe a course of treatment to Dr Ali for Sipho. In a matter of a few days Sipho was well again.

Case Study 7: *Apollo Hospital – using communications to advance health for rural society*

The Apollo Hospital group has established a state-of-the-art telemedicine facility at Aragobanda in Andhra Pradesh. This facility offers medical advice and support to rural communities and relies on ICT to fulfil its mandate. Through the telemedicine facility, medical specialists are linked with other medical practitioners and healthcare providers at remote clinics and hospitals to provide online real time support and facilitation for making medical diagnosis and recommending a course of treatment. This telemedicine facility supports the needs of about 50 000 people resident in Aragonda and six villages located in the region. The facility includes a fifty-bed hospital providing speciality in different areas, has X-ray and CT scan equipment and also includes an eight-bed intensive care unit and blood bank. In addition, it has high-tech equipment to scan, convert and communicate data images to the distant telestations situated at Hyderabad and Chennai. The facility also provides free health and screening services through camps set up for early detection of a variety of diseases. The telemedicine facility uses a very small apperture terminal (VSAT) to remotely connec to Hyderabad and Chennai. All rural families residing in the area of Hyderabad and Chennai areas have acess to the facility which costs around 0.25 US cents for a family of five persons.

Case Study 8: *Extending quality health services to the poor in Ginnack*

At the remote island village of Ginnack on the Gambia River, nurses are able to make use of digital cameras to take pictures of patient physical symptoms. The pictures are sent to a physician in a town nearby for examination. If in doubt of the didease, the doctor sends the pictures via Internet to the UK, where it is evaluated and examined by a medical institution with more knowledge. Compressed X-ray images are also sent across existing digital communications networks. In this way villagers of Ginnack are able to access better health care.

Case Study 9: *South African doctor's life saving SMSs*

A Cape Town doctor has dramatically helped the fight against tuberculosis (TB) by introducing a SMS service to remind patients to take their medication.

Dr. David Green, a consultant in Managed Care, Disease Management and Information Systems, became so frustrated when his mother constantly forgot to take her medication for hypertension that he started sending her SMS reminders – and it worked. Thanks to research he did for his PhD in Pharmacology, he was able to take his idea further and apply it to public healthcare. Dr. Green's reading eventually led him to two important insights. Firstly, he concluded that intervention designed to prevent non-compliance of treatment was not effective because it was applied indiscriminately. He concluded it was necessary to identify those patients who were non-compliant and find out why they were not taking their medication. Secondly, he was struck by the overwhelming literary evidence that suggested people were not taking their medication simply because they forgot.

It did not take him long to make the connection between the effectiveness of his SMS messages in alerting his mother, the high incidence of TB in Cape Town, and the possibilities that bulk SMS messages could present. However, when he wanted to pilot his innovative idea with TB patients at a local clinic in Cape Town, he met with resistance. Healthcare professionals were skeptical about the number of patients who would have access to cell phones. Not deterred, Dr. Green went back, did research and persuaded them with statistics that indicated that over 50% of Cape Town residents have cell phones. In addition, he found that 71% of patients at the clinic he earmarked for the pilot had access to cell phones. The local health authority eventually agreed and paid R11.80 per patient per month to run the SMS reminder service. The results of the pilot have been outstanding: of the 138 patients involved in the pilot, there was only one treatment failure.

The Medical Research Council of South Africa and the University of Cape Town have now embarked on a Randomised Control Trial to compare the cost-effectiveness of the SMS reminder service against the cost of non-compliance to TB treatment. In the mean time, the pilot has been so successful that the World Health Organisation has singled it out as example of best practice.

The initiative not only uses technology to address a real need effectively, but it does this in a simple, affordable and flexible way. Dr. Green uses a server, free software and a bulk SMS provider to send out the SMS messages. His system costs very little because he uses freely available open source software. In addition, his messaging system is flexible. When patients complained that the initial message ('Take your Rifafour now') was too drab, he added jokes, pearls of wisdom, and tips about lifestyle management to light up their day – but it still reminded them to take their medication.

Source: bridges.org[25]

These cases show that the impact of communications on developing nations is enormous. An important question that arises is: what role have businesses, governments, non government organisations (NGOs) and regulators been playing in taking communications to the mass market and are their efforts focused?

[25] http://www.bridges.org/case_studies/137 last accessed June 14, 2007.

BUSINESS, GOVERNMENT, NGOs AND REGULATORS – ARE EFFORTS FOCUSED?

The importance of communications as a tool for global integration and unification means that all parties have an important role to play in ensuring the equitable distribution of communications access for all. However, until early 2002, there was little incentive for enterprise or government to narrow the digital divide and efforts were not focused. The awakening of the industrial giants India and China has altered this. The wireless revolution sweeping across these countries has brought a rush of powerful supply chains aiming to target the bottom end of these markets. Their main aim is to get the population hooked to their mobile networks. Today, governments, non government organisations (NGOs) and other international donor agencies view the promise of forming wireless alliances with these supply chains as the best means for extending skills, health services and jobs to the rural poor.

The force, vigour and scale of the wireless revolution must never be underestimated. Global mobile penetration is estimated at around more than 1.5 billion phones and exceeds the 1.3 billion fixed line phones that took almost a hundred years to build. During the period before 2001, the global telecoms sector experienced a severe decline causing billions of dollars of debt. The introduction of mobile phone services in emerging markets caused resurgence in telecom growth and sector profitability in 2006. Generally, there has been strong growth in consumer communications spending globally and it is the one single sector that is growing faster than any other retail expenditure category. The universal center of mobile growth is Asia. Some 500 million or more new mobile phones per annum have been purchased in Asia since 2003.

Business efforts are focused at reducing prices for mobile phones so that they become affordable for more users at the lower end of the pyramid. Global ICT companies such as Motorolla, Sony Ericsson, Samsung and Nokia have initiated research and development centers, aiming to reduce the prices of mobile phones. At the same time, communications companies have introduced innovative affordable business models to make airtime more affordable. Examples of such innovations are prepaid airtime cards and the introduction of lower priced refurbished phones. In the majority of Asian and sub-Saharan developing countries, prepaid accounts for more than 95% of all airtime sold.

These innovative financing models have enabled a larger base of low income level users to secure access to communications platforms. Between 1999 and 2005, Chinese mobile communication connections surged from a mere 75 million to about 400 million. Ericsson[26] predicts that by 2007,

[26] Ericsson South Africa, 2006.

the number of mobile phone connections in China and India will be enormous. Chinese connections are expected to increase to more than 700 million while the Indian market is expected to increase by some 1 million or more connections per month and will grow to 55 million at the end of 2007.[27] This will mean a 5% penetration of the Indian market, which is relatively very small, compared to the population size of more than 1 billion. Mobile communications in Nigeria is estimated to be only about 18 million at the end of 2006 out of a total population of 130 million.[28] These figures may not appear to be much, but they are large enough to provide a base for business to consider ways of reaching out to mass markets.

The web of enterprise alliances and business ecosystems being formed today is insufficient to satisfy the broader long term goal of securing communications for all. There is a strong need to grow the web so that it becomes fully inclusive and extends beyond a handful of multinational business supply chains. The web must be woven in such a manner that it includes government, entrepreneurs, regulators, NGOs, academic institutions and anyone else that has a vested interest in achieving the long term objective. A new strategy is required that draws on the formation of a new class of wireless alliances that may work together to lower the costs so that communications becomes affordable to even the lowest income segments of the market.

CONCLUSION

In this chapter we looked at the impact of telecoms on developing markets. The significance of and need for technological progress in developing countries was evaluated and a number of important topics were discussed including the importance of information and information access for developing countries. We also looked at developing countries and the role of the Internet. The need for new sustainable business models to increase communications penetration in developing countries was also discussed. Thereafter, the correlation between telecommunications and economic growth was discussed and we then looked at the impact of communications in developing countries. Finally, we briefly looked at the role of regulators, enterprise and others and tried to determine if their efforts are focussed.

[27] www.indiaprwire.com last accessed on June 1, 2007.
[28] Ericsson South Africa estimates, 2006.

2

Mobile Telephony – A Great Success Story? Can Mobile Growth Be Sustained?

OVERVIEW

The mobile communications sector is one of the greatest business successes the world has ever seen. It took roughly twelve years to reach the first billion mobile users, and just two-and-a-half years to reach the second billion users and just one year to reach the third billion users. In fact, the global system for mobile conmmunication (GSM) has proven to be the fastest growing technology in history. The race continues and the industry is already focusing on reaching next milestone and that is 5 billion users by 2010. It started off a bit slowly in the early 1990s, but growth since then has been constantly accelerating. The idea of the first cellular network was brainstormed in 1947. It was intended for military purposes as a way of supplying troops with more advanced forms of communications. Between 1947 and 1979, several different forms of broadcasting technology emerged. A number of different analogue and digital mobile standards emerged across the globe. But GSM evolved as the only truly global standard that fuelled the growth of mobile communications.

The first radiotelephone service was introduced in the US at the end of the 1940s and was meant to connect mobile users in cars to the public fixed network. In the 1960s, a new system launched by Bell Systems, called Improved Mobile Telephone Service (IMTS), brought many improvements like direct dialling and higher bandwidth. The first

Business Models for Sustainable Telecoms Growth in Developing Economies
S. Kaul, F. Ali, S. Janakiram and B. Wattenström
© 2008 S. Kaul, F. Ali, S. Janakiram and B. Wattenström

analogue cellular systems were based on IMTS and developed in the late 1960s and early 1970s. The systems were 'cellular' because coverage areas were split into smaller areas or 'cells', each of which was served by a low power transmitter and receiver.

Mobile services based on GSM technology were first launched in Finland in 1991. Today, more than 750 mobile networks provide mobile voice and data services across over 220 countries and GSM represents over 84% of all global mobile connections. The invention of GSM technology is also attributed to the mass market phenomena for mobile communications. GSM's success is greatly attributed to an original vision of a cross border digital communications system, now used in almost every country of the world today. This vision enabled the development of a global ecosystem delivering vast economies of scale, massive product and service choice and global roaming, among other benefits for users.

China is the largest single GSM market in the world today followed by India, Russia and the USA; put together, these represent over $1/3^{rd}$ of global mobile customer based. In India, mobiles have become the fastest selling consumer product – pushing bicycles to the number two spot. The growth over last three years has been phenomenal. In 2007, India shall add over 80 million new subscribers, bringing the total to over 200 million. With the achievement of the 3 billion users milestone, GSM has become the first communications technology to have more users in the developing world than the developed world.

'What this means is that mobile phones are "bridging the digital divide" at an astonishing rate with relevant, affordable solutions that help families stay in touch, social networks to develop, businesses to grow and economies to develop.'

The telecoms industry has launched a number of initiatives to help people in the developing world gain access to mobile communications. As part of the International Telecommunications Union's (ITU) vision to 'Connect the Unconnected', the Emerging Market Handset initiative has delivered a first sub US$20 low cost mobile phone and efforts are being made to reduce the cost of handsets even further. Mobile manufacturers are putting lot of initiatives in place to create and develop innovative low cost & affordable mobile communications solutions that would prove attractive to a significant proportion of the world's unconnected people. For example, Ericsson one of the leading mobile infrastructure vendor and solution providers is driving an initiative called *Low ARPU High Growth*[1] which is focused on enabling communications for all and creating communications solutions that are suitable for driving profitability for service providers even at less than US$5 average revenue per user (ARPU) per month.

[1] Low ARPU High growth is an Ericsson group initiative that aims to brings mobile communication to every soul on earth.

The other hurdles that can be clearly identified are the tax barriers on mobile products and services in many emerging countries. The outcome of our desk research clearly demonstrated that taxes on mobile communications inhibit uptake and hence growth in mobile communication adoption, which in turn has an impact on economic growth and social development.

In tandem with growth of GSM in emerging markets, the take up of next generation 3GSM services globally is also flourishing. More than 120 networks across 60 countries have so far launched commercial 3G services. In Europe, where GSM is reaching saturation, some 95 percent of new connections are now for 3GSM based mobile services.

Today there are over 3 billion mobile subscribers globally and it is expected that there will be more than 5 billion subscribers by 2012, many of whom will be people earning even less than US$2 per day. The majority of new mobile users – around 80 per cent – will come from new high-growth markets underdeveloped and so-called developing nations of this which include countries like India, Russia, Pakistan, Brazil, Nigeria, the Democratic Republic of the Congo etc. While these users may have lower purchasing power on average than existing users, they represent a substantial growth opportunity thanks to their great numbers. What's more, studies have shown that mobile communication greatly increases individual income in developing, high-growth markets. Mobile communication has a positive impact both at a micro and macro level. At the micro level, mobile communication improves everyday life by increasing job opportunities, building stronger social networks, enhancing security and reducing the need for travel. At a macro level, mobile communication has a positive impact on society overall. It boosts efficiency, productivity, increases employment, enables development of rural areas and stimulates GDP growth.

ROLE OF GSM TECHNOLOGY

The Global System for Mobile communications (GSM: originally from Groupe Spécial Mobile) is the most popular standard for mobile phones in the world. The GSM service is used by over 3 billion people across more than 215 countries and territories. The ubiquity of the GSM standard makes international roaming very common between mobile phone operators, enabling subscribers to use their phones in many parts of the world. GSM differs significantly from its predecessors in that both signaling and speech channels are digital call quality, which means that it is considered a *second generation* (2G) mobile phone system. This fact has also meant that data communication was built into the system from the 3rd Generation Partnership Project (3GPP).

GSM is a cellular network, which means that mobile phones connect to it by searching for cells in the immediate vicinity. GSM networks operate in four different frequency ranges. Most GSM networks operate in the 900 MHz or 1800 MHz bands. Some countries in the Americas (including the United States and Canada) use the 850 MHz and 1900 MHz bands because the 900 and 1800 MHz frequency bands were already allocated. The less common 400 and 450 MHz frequency bands are assigned in some countries, notably Scandinavia, where these frequencies were previously used for first-generation systems.

In the 900 MHz band the uplink frequency band is 890–915 MHz, and the downlink frequency band is 935–960 MHz. This 25 MHz bandwidth is subdivided into 124 carrier frequency channels, each spaced 200 kHz apart. Time division multiplexing is used to allow eight full-rate or sixteen half-rate speech channels per radio frequency channel. There are eight radio timeslots (giving eight burst periods) grouped into what is called a TDMA frame. Half rate channels use alternate frames in the same timeslot. The channel data rate is 270.833 kbit/s, and the frame duration is 4.615 ms. The transmission power in the handset is limited to a maximum of 2 watts in GSM850/900 and 1 watt in GSM1800/1900.

GSM has used a variety of voice codecs to squeeze 3.1 kHz audio into between 6 and 13 kbit/s. Originally, two codecs (named after the types of data channel they were allocated) were used, called Full Rate (13 kbit/s) and Half Rate (6 kbit/s). These used a system based upon linear predictive coding (LPC). In addition to being efficient with bitrates, these codecs also made it easier to identify more important parts of the audio, allowing the air interface layer to prioritise and better protect these parts of the signal.

GSM was further enhanced in 1997 with the GSM-EFR (Enhanced Full Rate) codec, a 12.2 kbit/s codec that uses a full rate channel. Finally, with the development of Universal Mobile Telecommuniations System (UMTS), EFR was re-factored into a variable-rate codec called Adaptive Multi-Rate or AMR-Narrowband, which is high quality and robust against interference when used on full rate channels, and less robust but still relatively high quality when used in good radio conditions on half-rate channels.

There are four different cell sizes in a GSM network-macro, micro, pico and umbrella cells. The coverage area of each cell varies according to the implementation environment. Macro cells can be regarded as cells where the base station antenna is installed on a mast or a building above average roof top level. Micro cells are cells whose antenna height is under average rooftop level; they are typically used in urban areas. Pico cells are small cells whose diameter is a few dozen meters; they are mainly used indoors. Umbrella cells are used to cover shadowed regions of smaller cells and fill in gaps in coverage between those cells.

Cell horizontal radius varies depending on antenna height, antenna gain and propagation conditions from a couple of hundred meters to several tens of kilometers. The longest distance the GSM specification supports in practical use is 35 km or 22 miles. There are also several implementations of the concept of an extended cell, where the cell radius could be double or even more, depending on the antenna system, the type of terrain and the timing advance.

Indoor coverage is also supported by GSM and may be achieved by using an indoor pico-cell base station, or an indoor repeater with distributed indoor antennas fed through power splitters, to deliver the radio signals from an antenna outdoors to the separate indoor distributed antenna system. These are typically deployed when a lot of call capacity is needed indoors, for example in shopping centers or airports. However, this is not a prerequisite, since indoor coverage is also provided by in-building penetration of the radio signals from nearby cells.

The modulation used in GSM is Gaussian minimum shift keying (GMSK), a kind of continuous-phase frequency shift keying. In GMSK, the signal to be modulated onto the carrier is first smoothed with a Gaussian low-pass filter prior to being fed to a frequency modulator, which greatly reduces the interference to neighboring channels (adjacent channel interference).

A nearby GSM handset is usually the source of the 'di di di, di di di, di di di' signal that can be heard from time to time on home stereo systems, televisions, computers, and personal music devices. When these audio devices are in the near field of the GSM handset, the radio signal is strong enough that the solid-state amplifiers in the audio chain function as a detector. The clicking noise itself represents the power bursts that carry the TDMA signal. These signals have been known to interfere with other electronic devices, such as car stereos and portable audio players. This is a form of radio frequency interference (RFI), and could be mitigated or eliminated by use of additional shielding and/or bypass capacitors in these audio devices; however, the increased cost of doing so is difficult for a designer to justify. Figure 2.1 shows the structure of a GSM 2.5 network.

The unseen network behind the GSM system seen by the customer is large and complicated in order to provide all of the services which are required. It is divided into a number of sections and these are each covered in separate articles.

- The Base Station Subsystem (the base stations and their controllers).
- The Network and Switching Subsystem (the part of the network most similar to a fixed network). This is sometimes also just called the core network.
- The GPRS Core Network (the optional part which allows packet based Internet connections).

All of the elements in the system combine to produce many GSM services such as voice calls and SMS.

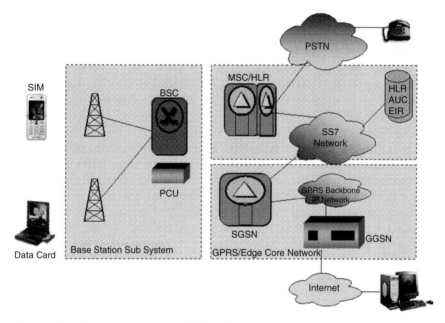

Figure 2.1 The structure of a GSM 2.5 network

Subscriber Identity Module

One of the key features of GSM is the Subscriber Identity Module (SIM), commonly known as a SIM card. The SIM is a detachable smart card containing the user's subscription information and phonebook. This allows the user to retain his or her information after switching handsets. Alternatively, the user can also change operators while retaining the handset simply by changing the SIM. Some operators will block this by allowing the phone to use only a single SIM, or only a SIM issued by them; this practice is known as SIM locking, and is illegal in some countries.

In the United States, Canada, Europe and Australia, many operators lock the mobiles they sell. This is done because the price of the mobile phone is typically subsidised with revenue from subscriptions, and operators want to try to avoid subsidising competitors' mobiles. A subscriber can usually contact the provider to remove the lock for a fee, utilise private services to remove the lock, or make use of ample software and websites available on the Internet to unlock the handset themselves. While most web sites offer the unlocking for a fee, some do it for free. The locking applies to the handset, identified by its International Mobile Equipment Identity (IMEI) number, not to the account (which is identified by the SIM card). It is always possible to switch to another (non-locked) handset if such a handset is available.

Some providers will unlock the phone for free if the customer has held an account for a certain time period. Third party unlocking services exist that are often quicker and lower cost than that of the operator. In most countries, removing the lock is legal. Cingular and T-Mobile provide free unlocking services to their customers after 3 months of subscription.

In countries like India, Pakistan, Indonesia, Belgium, etc., all phones are sold unlocked. However, in Belgium, it is unlawful for operators there to offer any form of subsidy on the phone's price. This was also the case in Finland until April 1, 2006, when selling subsidised combinations of handsets and accounts became legal, though operators have to unlock phones free of charge after a certain period (at most 24 months).

GSM Security

GSM was designed with a moderate level of security. The system was designed to authenticate the subscriber using shared-secret cryptography. Communications between the subscriber and the base station can be encrypted. The development of the Universal Mobile Telecommunications System (UMTS) introduces an optional Universal Subscriber Identity Module (USIM), that uses a longer authentication key to give greater security, as well as mutually authenticating the network and the user – whereas GSM only authenticated the user to the network (and not vice versa). The security model therefore offers confidentiality and authentication, but limited authorisation capabilities, and no non-repudiation.

GSM uses several cryptographic algorithms for security. The A5/1 and A5/2 stream ciphers are used for ensuring over-the-air voice privacy. A5/1 was developed first and is a stronger algorithm used within Europe and the United States; A5/2 is weaker and used in other countries. A large security advantage of GSM over earlier systems is that the Key, the crypto variable stored on the SIM card that is the key to any GSM ciphering algorithm, is never sent over the air interface. Serious weaknesses have been found in both algorithms and it is possible to break A5/2 in real time in a ciphertext-only attack. The system supports multiple algorithms so operators may replace that cipher with a stronger one.

Evolution

First generation: almost all of the systems from this generation were analog systems where voice was considered to be the main traffic. These systems could often be listened to by third parties. Some of the standards are NMT (Nordic Mobile Telephone), AMPS (Advanced Mobile Telephone System), Hicap (Nippon Telegraph and Telecom high-capacity System), CDPD (Cellular digital packet data), Mobitex, DataTac. 1G analogue system for mobile communications saw two key improvements during the 1970s: the invention of the microprocessor and the digitisation of the control link between the mobile phone and the cell site. AMPS (advanced mobile phone system), was first launched by US which is 1G mobile system. It is best on FDMA technology which allows users to make voice calls within one country.

2G (or 2-G) is short for second-generation wireless telephone technology. All the standards belonging to this generation are commercial-centric and they are digital in form. Around 60% of the current market is dominated by European standards. The second generation standards are GSM, iDEN (Integrated Digital Enhanced Network), D-AMPS (Digital Advenced Mobile telephone system), PDC (Personal Digital Cellular), CSD (Circuit Switched Data), PHS (Personal Handy-phone System), GPRS (General Packet Radio Service), HSCSD (High Speed Circuit Switched Data), and WiDEN (Wideband Integrated Dispatch Enhanced Network).

The main difference to previous mobile telephone systems, retrospectively dubbed 1G, is that the radio signals that 1G networks use are analogue, while 2G networks are digital. Note that both systems use digital signaling to connect the radio towers (which listen to the handsets) to the rest of the telephone system. 2G technologies can be divided into TDMA-based and code division multiple access (CDMA)-based standards depending on the type of multiplexing used. The main 2G standards are:

- GSM (TDMA-based), originally from Europe but used worldwide (Time Division Multiple Access)
- iDEN (TDMA-based), proprietary network used by Nextel in the United States and Telus Mobility in Canada
- IS-136 *aka* D-AMPS, (TDMA-based, commonly referred as simply TDMA in the US), used in the Americas
- IS-95 *aka* cdmaOne, (CDMA-based, commonly referred as simply CDMA in the US), used in the Americas and parts of Asia
- PDC (TDMA-based), used exclusively in Japan

2G services are frequently referred as Personal Communications Service, or PCS, in the United States.

2.5G is a stepping stone between 2G and 3G cellular wireless technologies. The term 'second and a half generation' is used to describe 2G-systems that have implemented a packet switched domain in addition to the circuit switched domain. It does not necessarily provide faster services because bundling of timeslots is used for circuit switched data services (HSCSD) as well. While the terms 2G and 3G are officially defined, 2.5G is not. It was invented for marketing purposes only. 2.5G provides some of the benefits of 3G (e.g. it is packet-switched) and can use some of the existing 2G infrastructure in GSM and CDMA networks. GPRS is a 2.5G technology used by GSM operators. Some protocols, such as EDGE for GSM and CDMA2000 1x-RTT for CDMA, can qualify as '3G' services (because they have a data rate of above 144 kbit/s), but are considered by most to be 2.5G services (or 2.75G which sounds even more sophisticated) because they are several times slower than true 3G services.

3G

3G is third-generation technology in the context of mobile phone standards. 3G describes the updating of cellular mobile telecommunications networks around the world to use new 3G technologies. This process is taking place over the period 1999 to 2010. Japan is the first country having introduced 3G nationally, and in Japan the transition to 3G was largely completed during 2005/2006. 3G technologies enable network operators to offer users a wider range of more advanced services while achieving greater network capacity through improved spectral efficiency.

The services associated with 3G include wide-area wireless voice telephony and broadband wireless data, all in a mobile environment. In marketing 3G services, video telephone has often been suggested as the killer application for 3G.

There were 170 3G networks in operation in 60 countries in the world. In Asia, Europe, Africa and the USA and Canada, telecommunication companies use W-CDMA technology with the support of around 100 terminal designs to operate 3G mobile networks.

Roll-out of 3G networks was delayed in some countries by the enormous costs of additional spectrum licensing fees. In many parts of the world 3G networks do not use the same radio frequencies as 2G, requiring mobile operators to build entirely new networks and license entirely new frequencies; a notable exception is the United States where carriers operate 3G service in the same frequencies as other services. The license fees in some European countries were particularly high, bolstered by initial excitement over 3G's potential. Other delays were as a result of the expenses related to upgrading equipment for the new systems.

The first country that introduced 3G on a large commercial scale was Japan. In 2005, over 40% of subscribers used 3G networks only, with 2G being on the way out. The successful introduction of 3G in Japan showed that video telephony was *not* the killer application for 3G networks after all. The usage of video telephony on 3G networks was found to be a small fraction of all services. On the other hand, downloading of music found strong acceptance by customers. Music download services in Japan were pioneered by KDDI corporation with the EZchakuuta and Chaku Uta Full services.

3G networks are not IEEE 802.11 networks. IEEE 802.11 networks are short range, higher-bandwidth (primarily) data networks, while 3G networks are *wide area cellular telephone networks* which evolved to incorporate high-speed internet access and video telephony.

The evolution of the 3G networks is gaining speed across the globe. The main reason for these changes is, basically, the limited capacity of the existing 2G networks to offer true multimedia services. The second generation of networks were built mainly for telephone calls and slow data transmission. Due to the rapid changes in technology, these factors do not meet the requirements of today's wireless revolution. The developments of so-called '2.5G' (or even 2.75G) technologies such as i-mode data services, camera phones, high-speed circuit-switched data (HSCSD) and GPRS have been ways of bridging the oncoming change to High speed networks, but are not permanent solutions. They are merely stepping stones towards the new technology.

The evolution of networks from the second generation of technologies to the third generation technologies could not be done without the help of network operators. In 2005 there were about 23 networks worldwide that operated on 3G technologies, today the numbers have exceed 100 networks, the most advanced being KDDI in Japan. Some of these networks were only for test use, but some were already in consumer use.

According to the 4G working groups, the infrastructure and the terminals will have almost all the standards from 2G to 3G implemented. The infrastructure will, however, only be packet based, all-IP. The system will also serve as an open platform where the new innovations can go with it. Some of the standards that paved the way for 4G systems are 3GPP Long Term Evolution, 3GPP2 Ultra Mobile Broadband, WiMax and WiBro.

4G is a short form of fourth-generation cellular communication system which provides end-to-end IP solution where voice, data and multimedia streaming can be served at higher data rates with anytime-anywhere concept. No formal definition is set to what 4G is, but the objectives that are predicted for 4G can be summarised in a single sentence:

4G will be a fully IP-based integrated *system of systems* and *network of networks* achieved after the convergence of wired and wireless networks as well as computer, consumer electronics, communication technology and several other convergences that will be capable of providing 100 Mbit/s and 1 Gbit/s, respectively, in outdoor and indoor environments with end-to-end QoS and high security, offering any kind of services anytime, anywhere, at affordable cost and one billing.

If you can dream it, you can do it, according to this we can leap 3G to 4G along its features and future trends in mobile technology. In wireless communication, mobile technology is advanced and in this system 4G is the latest at present. 4G mobile aims to provide an effective solution for the next generation mobile services. Progressing from previous three generations, 4G mobile systems have been significantly improving in terms of interactive multimedia services.

4G Technology enables:

- Interactive multimedia services: teleconferencing, wireless Internet, etc
- Wider bandwidths, higher bit rates
- Global mobility and service portability
- Low cost
- Scalability of mobile networks

Innovation and Mobile Growth

Growth of mobile phones is associated with firstly continuous innovation on behalf of telecoms industry, secondly that the development and the investment is mainly made by the private sector rather than governments or donor communities. The key milestones worth mentioning that catalysed the growth are as follows:

Prepaid: Innovative Pricing Scheme

The main explanations for this fast growth in mobile is invent of prepaid. Today if we take a look in developing economies, more than 90% of all mobile subscribers are prepaid. If we analyse why prepaid has been such success, there is only one argument that explains it and that is the miniaturisation of a service into a size that is affordable by the mass market.

If we just go decade back mobile services was primarily targeted at high-brow, high ARPU subscribers and the belief was that the poor would never opt for mobile services as against their basic needs like food and shelter. Communication was not perceived as the basic human need and that was and has been the fundamental stumbling block to mass market. The other reasons why it took over a decade to reach the first billion users and only two year to the next is attributed to fundamental barriers to uptake, these are tariffs and call price structures, cost of handsets, business case for bringing signals to rural and remote areas etc.

Village Phone Initiative and Microfinancing

Grameen Foundation USA and some mobile phone providers are collaborating to bring *'affordable, accessible telecommunications to rural villages in Africa through microfinance.'*[2]

[2] www.grameenfoundation.org/what_we_do/microfinence_in_action/faqs

The collaboration was built on GFUSA's global Village Phone initiative that is helping people living in rural areas start self-sustaining businesses while providing affordable Telecommunications to their communities. Access to these same affordable and reliable services is a lifeline for rural communities and a critical part of overall development. Although the costs of mobile phones and services have fallen significantly, the initial investment required for connection remains one of the most significant obstacles to expanding communication services in these areas. By providing the necessary cash, microfinance is proving a powerful tool in overcoming this barrier. What is more, with tiny loans, financial services and mobile technology, Village Phone provides affordable access in a sustainable manner.

Rural areas of Africa have been singled out by GFUSA so as to make universal access a reality in quicker than normal time. As part of their effort, GFUSA have jointly developed a solution based on low cost mobile phones and an external antenna to serve rural communities in Uganda and Rwanda, the two countries where GFUSA's Village Phone currently operates. Rural connectivity will play a major role in reaching the next few billion subscribers and ultimately connecting the majority of the world's population. By leveraging microfinance – a proven poverty reduction strategy and technology, GFUSA's Village Phone is playing an important role in alleviating poverty and connecting rural communities. GFUSA's Village Phone replicates the Village Phone programme that was pioneered in Bangladesh by Grameen Bank. It was launched in number of developing countries with great success.

MOBILE COMMUNICATIONS AND THE DIGITAL DIVIDE

The digital divide – the idea that, as information and communication technologies are affordable only to higher income groups, they in fact increase income disparities – has been at top of the agenda of any organisation concerned about socio-economic development. There has been a clear consensus for some time that bridging the digital divide goes a long way towards boosting development, and more recently the focus has been on building that bridge.[3] Bridging the digital divide depends on facing the affordability challenge in its three dimensions: total cost of ownership of mobile communications, the cash barrier faced by users to get and stay connected, and the regulatory environment.

[3] See www.ericsson.com/ericsson/press/releases/20070910-1152309.shtml and www.govtech.com/gt/141899

'Encouraging the spread of mobile communications is the most sensible and effective response to the digital divide.'[4]

Mobile phones bring a host of socio-economic benefits at both individual and national levels. Increased penetration and the additional productivity it generates at an individual level are passed on directly to accelerated growth in GDP.[5] A recent study by the London School of Economics showed that an increase of ten mobile phones per 100 people boosts GDP growth by 0.6 percentage points. Ericsson South Africa conducted qualitative research in countries in sub-Saharan Africa and found astonishing results, the correlation demonstrated the ratios as high as 1:4, which means every dollar spent in mobile infrastructure gave 4 dollars back to the economy. Also, the lower the GDP of the country, the higher was the correlation ratio.

Enabling universal access, that is, all people regardless of wealth and where they live should have access to communication and information via mobile communications is thus a noble goal and the target to aim for, as it will bridge the digital divide.

Mobile communications is playing a key role in eliminating the digital divide. Mobile phones do not rely on a consistent electrical supply and, unlike computers, can be used by people who cannot read or write. New business models are emerging that are paving a road for Communications for All.[6] For example, phones are widely shared and rented out by the call, for example, by the 'mobile ladies' found in some villages in Bangladesh. Mobile phones are used to make cashless payments in a number of African countries, so mobile phones are becoming a tool not only for communication but also for financial transactions and to conduct commerce.

Nigerian Tale: Creating a Mobile Payphone

One thing that unites the billions of GSM subscribers is the SIM. Since its inception it has been there in the same form factor providing network access, secure personal data storage and standardised presentation of operator services. As the next three billion users, mainly in emerging markets, become equipped, the SIM will be with every one of them. Mobile telephony has helped to link communities where until recently it had been unfeasible to install fixed line infrastructure. One of the

[4] www.ericsson.com/technology/whitepapers/can_mobile_communcations_close_digital%20divide.pdf
[5] www.delvelopingtelecoms.com/content/view/423/100 last accessed July 2007.
[6] Communication for all, an Ericsson global initaitive, for details browse http://www.ericsson.com/drivingthemarket/comsforall/index.shtml

by-products of the GSM boom has been the use of cell phones as payphones.

This is helping to reach customers who previously could not afford a mobile phone. For example, of the 140 million people in Nigeria roughly 30 million of them are GSM users, whereas it is felt that nearly 64 million more are potential payphone customers.

While this has opened telephone use to the masses it has its flaws. Many potential customers cannot afford the minimum call rates and disputes are frequent between customer and reseller over the call time. Again in Nigeria, the telephone market traditionally uses stopwatches in the streets to time customers. Because the minute doesn't calculate correctly and the phone doesn't cut off, there have been disputes. Owners of public payphone businesses in Nigeria cannot control what is being spent on the phones that they have put in the market. A SharedPhone SIM allows a street operator to use a normal GSM handset as a public payphone.

The SIM application lets the reseller set the minimum charge (in Nigeria this went from 30 Naira (23 US$ cents) to five Naira (4 US$ cents), fix call duration and accurately display the call timing. The result is that more customers can afford to make calls while they and the reseller have increased confidence in the charging system.

In many developing nation across the globe, SIM-based airtime reselling is giving many millions of people access to telephony for the first time while encouraging small business growth. Payphone operators on the streets of developing countries are now able to set their own billing tariffs and manage and grow their payphone businesses at a fraction of the standard price.

Many people are setting up their own payphone business as the main source of income, but it is also proving successful with people who have existing businesses, a perfect example is a taxi owner or someone who runs a hairdressing salon. Most taxi owners or drivers do have a phone, but they could also have a shared payphone because they can double their income or on a smaller scale and make 20 to 30% more on a monthly basis by having the application. Following the initial deployments in Nigeria, SharedPhone[7] looks set to roll-out in Mozambique, Zimbabwe, Lesotho, Uganda, the DRC and South Africa. As in other emerging markets around the world, African operators have seized the opportunity of offering their subscribers mobile services and see the SIM as a way of targeting as many people of possible. Their customers are therefore benefiting from increased voice and data communications while mobile payphones are helping to stimulate small business growth.

[7] SharedPhone: SharedPhone offers a SIM based GSM Public Payphone for local entrepreneurs who would use it to provide phone services, details on www.next-billion.net/activitycapsule.

Even though the number of phones per capita in many emerging markets is much lower than in the developed world, they can have a dramatic impact: benefits include reducing transaction costs, broadening trade networks and reducing the need to travel. As a result, consumers in emerging countries often spend a larger proportion of their income on telecommunications than those in more developed ones.

Mobile Phones and Their Positive Impact

Our detailed analysis of micromarket data in Nigeria shows how phones really do make people better off; and best part is that mobile phones do this without the need for government intervention. Mobile phone networks are built by private companies, not governments or charities, and are economically self-sustaining. Mobile operators build and run them because they make a profit doing so, and fishermen, carpenters, porters and alike are willing to pay for the service because it increases their income. The resulting welfare gains are indicated by the profitability of both the operators and the mobile users. All governments have to do is to issue licenses to operators, establish a clear and transparent regulatory framework and then wait for the phones to work their economic magic.

ENRICHING COMMUNICATION BETWEEN PEOPLE

Communication is a basic human need and industry is moving into a new era in which mobile communications is playing an increasing role improving quality of human lives. Communication culture is changing. Communication has moved from making announcements to sharing everyday experiences, thoughts and emotions as well as self-created or purchased content.

The youth are setting the scene: already accustomed to having broadband and communications all around them, they are fast to adopt new, richer communication services.[8] For example, a study of 10 major countries in Europe, the Americas and Asia shows that 85 percent 15-to-24-year-olds never leave home without their phones and 31% use instant messaging every day. Richer communication is gaining ground simply because it enables individuals to fulfill the important needs for closeness, convenience and control.

---

[8] Ericsson Consumer Lab Study, 2006.

Mobile Phones are Bringing People and Communities Closer

Closeness means staying in touch with family while away, bringing relatives from different locations together and being part of an event or gathering when physical presence is not possible. Allowing users to reach out across distance and allowing users to share more about where they are, what they are doing and what is going on at any given moment will help them achieve a greater sense of closeness.

Control of your communication situation and of how you are perceived is also an important benefit. Individuals want the security of knowing 'someone is always there,' for support, for help with decision-making, or for help in an emergency.

Convenience is achieved through more flexible and spontaneous ways for people to visualise what they are talking about. People want to share objects and ideas, show images and videos and hear sounds without having to interrupt their conversations.

RICHER COMMUNICATION AND SUSTAINABLE SOCIETY

By fulfilling these needs for closeness, convenience and control, richer communication can also contribute greatly towards a more sustainable society.

Consider the following examples:

- Bringing people closer virtually rather than physically limits travel needs, reducing energy consumption and greenhouse gas emissions.
- By introducing visual elements in telecommunications, the hearing- and speech-impaired can communicate much more easily, improving their lives substantially.

More control also makes it possible to more speedily and efficiently provide medical services and critical information to those in need, regardless of physical distances. Advancements in communications can be seen as enabling technologies for sustainability to become a reality, providing undisputed social benefits and a toolkit for environmental protection.

The number of mobile users is growing twice as fast in developing countries as in developed countries. Africa is now the fastest-growing mobile market in the world. The rapid spread of mobiles has been aided by prepay options that allow users to control their spending. The number of mobile users is often much higher than the actual number of phones, as many people allow family and friends to use their phones.

Community phone shops allow many more people to gain access to telecommunications.

Increased mobile connectivity improves access to information. Knowledge of latest prices in different markets, for example, can improve price transparency for small farmers and fishermen who can cut out the middlemen and gain direct access to markets. For example, one of the leading mobile operators in Kenya, Safaricom, helps farmers keep track of market prices for their goods.

Case Study 1: *Helping farmers obtain a fair price for their produce in Kenya*

Safaricom has introduced a text messaging service that provides quick and easy access to updates on agricultural markets. Buyers and sellers of agricultural commodities can keep track of prices using the Sokoni Short Messaging Service (SMS) on their Safaricom mobile phone. This helps users obtain a fair price for their goods.

The Sokoni service transmits daily reports from the Kenya Agricultural Commodity Exchange (KACE). Users simply text the name of the commodity they are interested in, such as 'maize' or 'sheep', and receive an instant reply with an update on its price that morning at each market. This enables them to assess the best place and time to buy or sell. KACE is an NGO that helps to link farmers and traders and provides daily reports on commodity prices from all major Kenyan markets. By using the Sokoni service, traders and farmers can get market updates on the day, rather than waiting for the next day's newspaper. The service also allows traders to offer their goods for sale or place bids, as well as post short messages or agricultural questions.

MOBILE COMMUNICATIONS VERSUS SOCIAL ECONOMIC DEVELOPMENT

The impact of mobile phones extends beyond measurable economic indicators. Much attention has been focused recently on the digital divide, which refers to the fact that in many parts of the world income differences are reflected in a gap between those with and those without access to computing and communications technology. This divide is regarded as responsible for further isolating the world's poor. Various initiatives have been undertaken to help bridge this divide, including the World Summit on the Information Society, endorsed by the UN.31 Experience has shown that regional centres equipped with computers providing internet connectivity have had a very limited impact. This is because of a lack of literacy, the distance of centres from large parts of the population, and technical problems, especially in relation to the maintenance of these centres. Mobile technology, on the other hand, has proved extremely effective in addressing the issue of the digital divide. Literacy is not required, the mobility of the device helps deliver versatile services (wherever there is coverage) and handsets are robust and easy to use. This explains why it has been suggested the spread of mobile technology would be the most effective way to address the digital divide.

Benefits Brought to Poor Communities

There are many examples that suggest that providing mobile connectivity in poor communities will indeed deliver significant social benefits. The importance of mobile coverage to communities is starkly illustrated in a case mentioned to us by a leading operator in sub-saharan Africa. After building a mobile phone mast in the countryside of one African country, the operator returned to the site a couple of months later to find that a nearby village had relocated next to the mast where mobile phone reception was best. On later visits, electricity and water supply had become available, and a road had been built to the site. To the extent that mobile phones facilitate attempts to create successful small businesses and help poor individuals to find work, they will also help to reduce inequality within countries. Naturally, this will depend on the extent to which services and coverage can be provided to poorer communities. Since many of these communities are located in rural areas away from big cities, this presents a challenge to mobile operators and regulators. Careful management of incentives and universal service obligations will be critical in this respect. The social impact of mobile networks is much higher in countries where no extensive fixed network exists (as is the case in most of sub-Saharan Africa). In many cases, mobile communication is the only way of receiving information on the availability of medical treatment in regional centres. Prior to the introduction of mobile telephony, for many families regular communication between rural and urban relatives was impossible. In emergencies, travel was the only way to convey important messages. In this context, mobile services have a profound influence on people's lives. Even if ownership of a separate subscription is beyond the means of poorer relatives, many people are finding innovative ways of staying in touch. This may involve maintaining contact with another member of the community, who does own a mobile phone, or traveling to a nearby village or town where mobile payphone services are available. In this respect, community service phones have made significant progress in extending access to poor and rural users. A survey carried out for the UNDP by Ericsson found that 97% of the population in the villages surveyed already knew about mobile telephony, and 50% had used a mobile phone. By contrast, only 33% knew what a computer was and only 3% had ever used one.[9]

Methods to ensure that wealthier friends and family members bear most of the costs of mobile communication are also common, for example, the practice of 'beeping' whereby the caller hangs up before the other party picks up. In this way, people send a signal that they wish to be contacted without incurring any cost. Certain operators in the region have taken note of this communication need and offer services whereby the call costs are borne by the party that can better afford them. This includes services such as the free 'call-me' SMS.

More Benefits and More Evidence

In developing countries, mobile networks provide many people with access to new activities outside their traditional sphere. The absence of fixed networks means that mobile connections often offer the only opportunity for Internet access.[10] The World Bank has pointed out the benefits of the Internet for developing countries, including increasing trade between regions, developing rural agricultural markets, healthcare delivery, and HIV/Aids prevention.

A recent development has been the introduction of *m-Commerce* financial services, which phone users can use for purposes such as making cash deposits and withdrawals, transferring money between user accounts, and making retail purchases. Such systems have been applied most successfully in the Philippines, but services have also been launched in several countries in Southern Africa.

The value of communication to developing country users is illustrated by the willingness of low income users to pay for services. In interviews

[9] www.ericsson.com/drivingthemerket/doc/communication_for_all.pdf
[10] en.wikipedia.org/wiki/World_Bank

with stakeholders it was estimated that many users in sub-Saharan African spent 8–10% of their income on communication services. For the lowest income groups, expenditure constitutes an even more significant share of income. Souter et al.,[9] found that some users in Tanzania spend around 14% of their income on telephony.

Examples:

Nigerian Hospitals[11]

A study of ICT (Information and Communications Technology) use in Nigerian teaching hospitals found that 99% of doctors used mobile phones.[12] This had a direct impact on the quality of medical services provided and the health of the patients within the hospitals. Doctors used mobile phones for communications between wards, including referring patients on to other practitioners, contacting colleagues for consultation or second opinions and arranging for equipment or materials to be brought to a particular ward. Although mobile phones had a clear impact on the level of services, doctors were bearing the entire costs themselves including purchasing handsets and call charges.

Ugandan Village

The Village Phone Uganda project, sponsored by the Grameen Foundation and MTN Uganda, has been supporting the extension of cell-phone coverage in rural areas. The project offers microfinance to allow female entrepreneurs to purchase a mobile phone. Phone services are then sold on to other members of the community. Village Phone operators are offered a starter kit, which consists of a mobile phone, a SIM card, prepaid phone time, business cards, a sign and a car battery or solar panel to recharge the phone. Clients use the phones to conduct business transactions, communicate with family members, check the prices of agricultural goods and to participate in radio call-ins. Access to mobile communications drives economic growth, improves business, education and employment opportunities and enables people to keep in touch with family and friends. Many people are denied this access, particularly in developing countries.

Mobile phones are enabling tens of millions of people with no access to postal or fixed telephone services to become connected to the wider

[11] Journal of Information Technology Impact 3(2), 2003.
[12] www.portal.acm.org/citation.cfm?id=1159435

world. In 19 African countries, mobiles account for three quarters or more of all telephones, and the proportion is even higher in a few countries, such as Benin and Kenya.

A number of recent studies by Ericssons' Consumer LAB analysing the socio-economic impact of mobile phones (SIM) are helping to inform the debate on the digital divide. The SIM research shows that mobile phones play a significant part in promoting bottom-up socio-economic development, even among the poorest communities.

Access to mobile telecommunications in developing countries can help bridge the digital divide. Mobiles do not require the same levels of education and literacy as other new technologies such as computers or the Internet. This makes them more accessible. Accessibility is further increased by the lower up-front expenditure required and flexible pricing plans.

Mobile communications makes a direct economic contribution through job creation, taxes, payments to suppliers, dividends to shareholders and returns to lenders.

Findings from the Asian Continent:

Mobile Phones Promote Economic Growth for Fishermen in Northern Kerala, India

You are a fisherman off the coast of northern Kerala, a region in the south of India. Visiting your usual fishing ground, you bring in an unusually good catch of sardines. That means other fishermen in the area will probably have done well too, so there will be plenty of supply at the local beach market: prices will be low, and you may not even be able to sell your catch. Should you head for the usual market anyway, or should you go down the coast in the hope that fishermen in that area will not have done so well and your fish will fetch a better price? If you make the wrong choice you cannot visit another market because fuel is costly and each market is open for only a couple of hours before dawn and it takes that long for your boat to travel from one to the next. Since fish are perishable, any that cannot be sold will have to be dumped into the sea. This, in a nutshell, was the situation facing Kerala's fishermen until 1997. The result was far from ideal for both fishermen and their customers. In practice, fishermen chose to stick with their home markets all the time. This was wasteful because when a particular market is oversupplied, fish are thrown away, even though there may be buyers for them a little farther along the coast. On average, 5–8% of the total catch was wasted. But starting in 1997 mobile phones were introduced in Kerala. Since coverage spread gradually, this provided an ideal way to gauge the effect of mobile phones on the fishermen's behaviors, the price of fish and the amount of waste. For many years, anecdotes have abounded about the ways in which mobile phones promote more efficient markets and encourage economic activity. One particularly popular tale is that of the fisherman who is able to call several nearby markets from his boat to establish where his catch will fetch the highest price.

As phone coverage spread between 1997 and 2000, fishermen started to buy phones and use them to call coastal markets while still at sea. (The area of coverage reaches 20–25 km off the coast.) Instead of selling their fish at beach auctions, the fishermen would call around to find the best price. Dividing the coast into three regions, it was found that the proportion of fishermen who ventured beyond their home markets to sell their catches jumped from zero to around 35% as soon as coverage became available in each region. At that point, no fish were wasted and the variation in prices fell dramatically. Waste had been eliminated and the 'law of one price' (the idea that in an efficient market identical goods should cost the same) had come into effect, in the form of a single rate for sardines along the coast. This more efficient market benefited everyone. Fishermen's profits rose by 8% on average and consumer prices fell by 4% on average (SIM research conducted by Ericsson in 2007). Higher profits meant the phones typically paid for themselves within two months. And the benefits are enduring, rather than one-off. All of this shows the importance of the free flow of information to ensure that markets work efficiently. Information makes markets work, and markets improve welfare.

Findings from the African Continent

Felix lives with his parents and younger sisters in a village outside Benin, Nigeria. He is a farmer and, as the oldest son, is expected to take over the farm and the house one day. The family lives off the money they make from the farm, but because this is not enough to provide for food and education, Felix's younger brother, who works as a trader in Lagos, sends them money each month.

The process of delivering the money is difficult and costly. A friend of the brother brings the money when he travels to Benin to see his family each month. He is paid 400 naira (US$3.25) if he takes the money as far as Benin and 1000 naira if he brings it out to the village.

The friend sometimes delivers the money to the village, but most of the time Felix has to pick it up in Benin, a trip that takes one or two hours by bus and costs 400 naira for a return ticket. In Benin, Felix makes a call to the friend from a local business centre and then waits for him to arrive with the money. If the friend is busy, Felix sometimes has to wait a long time. It is difficult to live on US$2–5 a day. Every opportunity to make a few extra cents is crucial. In the rural areas, people usually earn their small and fluctuating incomes by cultivating small plots of land and selling a couple of litres of milk. People may also have very small businesses (kiosk, cobbler, blacksmith or tailor) or take on casual work (picking tea, chopping wood).

People who earn more than US$5 per day tend to be employed and have a business on the side. The major difficulty in the city, as in rural low-income segments, relates to making money. Apart from food costs and school fees, people must pay rent for housing as well as electricity and water bills. Money is a constant consideration as people decide how to make use of limited resources, perhaps being forced to decide between whether to take the bus or eat lunch. A typical rural area consists of one main village surrounded by several smaller villages. Many of the common amenities (school, court, health centre) are located in the main village, where about 600–1000 people live. The smaller villages are 4–10 km away from the main village. Each village is made up of privately owned houses where the villagers live in close proximity to each other. Farms are located outside the villages. Some farmers have to walk two or three hours to reach their land. Many men leave the village to find jobs in a city to support their families back home. This is the reality for people living in rural areas of most developing economies.

Understanding Real Life Economics

Mobile phones satisfy basic needs by generating the extra income so crucial for low-income households to make ends meet. The telecom business community thus needs to gain deeper insights into how communication changes the lives of the poor and what their priorities are.

> 'It is difficult for us because we don't have much money. I try to help on other people's farms but they don't always need me. If phone calls were possible, I could find jobs in other villages, too.'
>
> GRACE, 36, ODUNA, NIGERIA

Does life really change with mobile phone coverage? And if so how? How can people living on US$2–5 a day benefit from mobile telephony? The story about Felix comes from our research in Nigeria last year. Now let's look at a similar story, with the same size transaction, for someone who has a mobile phone and lives in an area with coverage.

Ronke has a sister who attends school and is often short of cash. When Ronke's sister has run out of money, she calls or texts Ronke, who then sends an SMS with a top-up card number. The sister sells the top-up card at a business centre (for a fee), converting it into cash.

So instead of paying more than 850 naria for someone to bring the money to a nearby city, plus bus expenses, waiting time and so on, Ronke's sister pays a fixed fee of 100 naria to convert airtime into cash. This is a saving of at least 750 naria and about four hours of time, plus there is no risk of being tricked by the person who is bringing the money – a dramatic change made possible by a mobile phone and the creativity of people on low incomes. Our studies in Kenya and Nigeria reveal that mobile networks really can enhance living conditions for poor people. Mobile phones enable major developments in business opportunities and the ability to maintain vital social networks. These effects are evident for the people, the villages, and the cities, as well as for the countries as a whole.

Having a mobile phone has made it much easier for Amina and her family to get health assistance when someone in the family is sick. With three small children in the household, Amina often needs to get in touch with a nurse or doctor. With a mobile phone, the family can get advice by making just one call. If an examination is needed, they can arrange for the doctor to make a house call, or they can make an appointment to see the doctor at the health centre. Previously, Amina's husband had to walk 4 km to the nearest health center without knowing if the doctor was actually there. Infrastructure in these countries has often been neglected for years, resulting in frequent power cuts, bad roads and a lack of public transportation and public phones. Getting in contact with people living beyond walking distance is difficult, as traveling is associated with long journey times and danger. Mobile phones reduce stress and risk, and save time, connecting people in a fast, easy and relatively cheap way. People often say about the mobile network: finally, a modern infrastructure that works.

Mobile Phones Create New Business

Hasia received a mobile phone from her sister who lives in Saudi Arabia. This as a great gift because Hasia can use it to earn extra money for the family. Each day, Hasia sends her younger brother out to the main road with the mobile phone, where he sells phone calls, bringing in an extra 1500 naira per month for the family. Hasia can't do this herself because she is not allowed to go out, except when her husband gives her permission to visit close friends or family.

When earnings are limited, calls are made strictly in relation to earning money. When income increases, calls to family quickly become more frequent and longer. With higher income, most calls are to family, although calls to friends also increase. One reason is that people are employed to a larger extent and have less need to make the calls they previously made in the running of their small businesses.

Another important and related finding in the research is that the proportion of income spent on mobile phone calls is much higher among those with lower levels of income. (Ericsson SIM Research 2007)

In addition to increased earnings for ordinary businesses, the network itself opens up the possibility for new sources of income. Enterprising people in rural areas can start up their own businesses without having to move to the city and city dwellers can make a living from mobile-related businesses.

Private Calls Strengthen Family Networks

The absence of a social security system means poor people have to rely on family to survive. Mobile phones as a means of communication are invaluable for maintaining family networks and calling for assistance. With no mobile network, it may take weeks before somebody can respond to a cry for help. The mobile network has made a dramatic difference. Communications have become easier, enabling all sorts of information to be passed on more easily.

'We didn't hear from my brother for three months, so we all agreed that I should go to see him. He had not been paid his salary for many months. That was why he hadn't sent a messenger.'

FELIX, 38, ODUNA, NIGERIA

'I was really happy when my sister gave me a mobile phone. I soon realised that I could ask my brother to go to the roadside to sell calls and that has given us some extra money, which we really need.'

HASIA, 22, JOGANA, NIGERIA

Unemployment is high and many people rely on the informal sector for their incomes. Competition is fierce as many people sell the same products; those who run their businesses more efficiently are therefore more likely to succeed. In an environment where every deal is crucial, mobile phones have made it possible to negotiate prices and arrange deliveries from suppliers, as well as to build and maintain customer relationships, resulting in more business and increased income.

'We were planning to buy a goat, but decided to buy a phone instead for our business ... Life is cheaper when I have the mobile phone and don't have to pay for the matatu[13] (minibus) to Muranga.'

NANCY, 33, KANYANAINI, KENYA

[13] Matatu stands for minibus in Sohali, matatus are used as the public transport in Kenya.

Several self-employed men and women in the rural villages and low-income areas of these countries run small-scale enterprises. Mobile phones make it possible to create and maintain customer relationships.

"I can actually earn money by using my mobile phone. That's why most of my calls are business related. I know that even if I am going to spend 100 (Kenyan) shillings (US$1.45), at the end of the day it is going to give me 1000 shillings back."
ROGERS, 37, NAIROBI, KENYA

There is a correlation between income and communication habits. Increased income and a desire to solve logistics issues are seen as the main drivers for using a mobile phone. A phone can mean a new life.

- **'Umbrella' operators:** it is estimated that 600 000 people are engaged in selling airtime in Nigeria today.
- **Repairs:** self-employment for young men with technical qualifications and creative minds.
- **Business centers:** self-employment opportunities for people who might otherwise be idle.
- **Top-up cards:** new, increased revenues for shops, kiosks and other outlets, selling top-up cards, as well as income opportunities for people selling cards in the traffic.
- **Retailers:** sales of new and second-hand mobile phones.
- **General greetings**: finding out how people are. There are three main reasons for making calls.
- **Family matters**: all kinds of family-related news including births, weddings, funerals, financial difficulties or problems with drugs, school or work.
- **Support**: asking for advice or money. Low-income segments are used mainly to discuss important issues concerning monetary assistance, investments, and private family matters, privacy are a great concern. One advantage of having access to a mobile handset is that it offers privacy and enables direct contact – benefits that are unavailable when using a business centre phone.

Pricing Crucial for Take-Up

The single most important barrier to the continuous spread of mobile phone ownership and usage is money. Extremely poor households (earning less than US$2 per day) will have problems finding the means to buy a handset, although there are cheap secondhand models for sale. For households earning US$3 and more, a mobile phone is within reach and families may make joint efforts over a period of time to save up money to buy one. Some refrain from buying new clothes, others buy really cheap food or postpone purchases such as a TV or radio. Another option is to halt payments to parents back in the village. Even when they have mobile phones, low-income mobile phone users continue to use business centers and 'umbrella' operators because they have no credit on their mobile phones and cannot afford to buy the cheapest top-up card; it is cheaper to call from a business centre than to make a mobile phone call to another network; and it is cheaper to make international calls from a business centre than from a mobile phone. Use of Mobile phones dramatically improves living conditions for poor people and increases their belief in a future for themselves and their families. Improved living conditions and a greater belief in the future result in reduced migration to the cities. A mobile network connects villagers with cities and with the rest of the world, making it possible to run efficient businesses in village regions, and to maintain social networks while still living in the village.

Saving Outweigh Call Cost

Puriti's family has experienced a drop in income since her husband was let go from his job at Barclays Bank in Nairobi two years ago. Puriti's husband is still in Nairobi doing casual work, while Puriti and the children live in Kanyanaini, where Puriti works as a teacher and takes care of their animals and small field. They earn some extra money selling tea and eggs. Puriti also tries to save money by using a mobile phone to order fertilizer for the field. Fertilizer in the larger town of Kangema is 50 shillings cheaper per bag than in Kanyanaini. Puriti therefore calls a distributor in Kangema when she wants to place an order and the goods are delivered to her door. Puriti saves 50 shillings per bag when she buys five bags. This leads to a saving of 200 shillings – the phone call usually costs 50 shillings.

Mobiles are used and owned differently in developing countries from the developed world. The value of mobile phones to the individual is greater because other forms of communication (such as postal systems, roads and fixed-line phones) are often poor. Mobiles provide a point of contact and enable users to participate in the economic system. Many people who cannot afford to own a mobile themselves can access mobile services through informal sharing with family and friends or through community phone shops.

Use of text messaging in rural communities is much lower due to illiteracy and the many indigenous languages. This has implications for other technologies that use the written word, such as the Internet.

Mobile Communication Has Improved Economic Growth, Quality of Life and Social Capital

Mobiles have a positive and significant impact on economic growth. This impact may be multifold in developing countries as in developed countries; lower the GDP of the country higher the impact. A developing country with an extra 10 phones per 100 people between 2000 and 2006 would have had GDP growth 1.3% higher than an otherwise identical country (Ericsson SIM Research 2007).

Fixed and mobile communications networks, (in addition to the openness of the economy, the level of GDP and other infrastructure), are positively linked with Foreign Direct Investment. The impact of mobile telecommunications has grown in recent years.

Many of the small businesses surveyed use mobile phones as their only means of communication. For example 40% of the small business surveyed in four metros of India, roughly 60% of the small businesses surveyed in South Africa and almost same in Kenya Nigeria and Egypt said they had increased profits as a result of mobile phones, in spite of increased call costs being high in South Africa and Egypt.

Mobiles are used as a community amenity. Most mobile owners surveyed in Kenya allow family members, friends and business associates to use their handset for free and a third do the same for friends.

Case Study 2: *Living on the Edge*[14]

There is increasing evidence which suggests that mobile phones represent the most wide-reaching ICT (Information and Communication Technology) solution which can narrow the digital divide and bring growth to the most underdeveloped parts of our world. The ICT sector has enormous capacity to make basic and sustainable changes to the lives of people in third world countries. However, despite recent development more than two thirds of the world's population does not have affordable access to either voice or data communication.

Main Findings

Following are the main finding of our consumer research across sample developing economies:

- Low income households provide a great market potential for service providers
- The mobile phone satisfies basic needs (rather than desires) by generating the extra amount of income that is so crucial for low income households to make ends meet.

---

[14] Living on the edge, is a research study conducted by Ericsson Consumer Lab in Kenya and Nigeria, presenting down to earth and compelling examples of how Mobile communication is making lives better and richer at the lower end of the consumer pyramid; Augur Marknadsanalys.

- Increased income and efficiency in very small businesses is the main driving force for buying a mobile phone within low income segments.
- The mobile phone enables direct contact with suppliers and customers and eliminates travel and waiting time which makes the sale and supply process more effective – resulting in increased income on a micro level and probably also increased GDP on a macro level.
- A mobile phone is affordable for households with a daily income of US$3, whereas households with US$1 per day or less cannot, in most cases, afford to buy and use a mobile phone.
- Lower denominations of top-up cards and cheaper handsets would make mobile phones even more accessible for low income segments.
- Successful transfers of airtime and information from institutions via SMS (such as electrical bills and bank accounts) opens up for further development of financial services via the mobile phone.
- Computers are very scarce and young people are keen on using the mobile phone for data applications. However price is definitely a huge barrier among low income segments.

Living on US$1 Per Day

In the rural areas visited, families live in small houses (of 2–3 rooms) built from corrugated tin, wood or clay, with a trampled dirt floor and a few pieces of furniture. There is usually a separate shed for cooking facilities, although there is no electricity or running water (in some villages there is a tap in the compound).

Everyone owns chickens and most people also own a cow. Households normally consist of 3–6 people. Young couples often build their own houses next to the man's parents' plot, as the plots or *shambas* are divided and inherited.

Small and fluctuating incomes are earned by people through the cultivation of their own small shamba, and through selling a couple of litres of milk. People also possibly have a very small business (kiosk, cobbler, blacksmith, tailor) or take on casual work (picking tea, chopping wood, driving the matatu bus). Many men leave the village to find jobs in a city to support the family back home. They regularly come back and visit.

Julie and George live together with their two children on a small shamba where they have built a one-bedroom house constructed from wood with a corrugated tin roof. The household has two sources of income: Julie takes care of the shamba and sells tea leaves and milk, while George runs a small business as a blacksmith in the commercial town centre, 5 km away from where they live.

In the morning, the family all wake at around 5.30 am, get dressed and have something to eat. Julie has already milked the cow and George heads off to the town centre and his business. When Julie has waved the children off (they walk the 4 km long route to primary school), she carries the milk to the roadside 2 km away, where the milk float will come and pick it up. Returning to the shamba, Julie starts picking tea leaves and, when finished, carries the basket of leaves to the town centre where the tea truck comes and picks it up. Julie normally makes it in time for the 10am–12 noon pick-up slot. If she has money she might buy sugar, flour or cooking fat in the town centre; otherwise, she chats a little with friends and walks back home to the shamba. Back at home, she cleans up or washes clothes, takes care of the eggs (from the family's own hens) and picks maize, beans or bananas grown for the family's own consumption. She also milks the cow again as the milk will be used by the family in the afternoon. When the children return home from school they help Julie with the household work. In the evening she makes dinner and puts the children to bed, before going to bed herself.

When George has opened his shop, a neighboring businessman drops by and chats with him for a while. The neighbor mentions that business has been slow of late, and George agrees. Nobody has placed an order for a long time now and income has been low. This is not uncommon: George is used to business going up and down and to not knowing when he will next receive some money. George's mobile phone rings and it is a customer who wants to buy security bars for his windows. The terms of the deal are discussed and, as the customer finds everything satisfactory, he agrees to stop by the shop and decide on the design of the bars. When everything is settled George starts working on the security bars. He realises that he has to order more supplies and at 6 pm uses his mobile phone to call his supplier and place an order. George tries to make most of his calls between 6 pm and 6 am because of the off-peak tariffs. The goods are in stock and will be delivered a couple of days later. George then closes the shop and stops by the bar, where he chats with some friends before heading home for dinner.

Urban Life in Suburbs of Nairobi

There are notable low income areas (estates) surrounding the hot and polluted central parts of Nairobi. Compact living is the norm, and families often share facilities (e.g. toilets, shower and kitchen) with neighbours. The run-down estates have been neglected for years and people feel inse-cure due to assaults and robberies. Many are striving to move to a better area and hope for an improved life in the future.

The composition of households differs greatly, varying between 1–6 people with a variety of reasons for living together (e.g. singles, friends, couples, families, relatives). Relatives moving to Nairobi from rural areas regularly visit for longer periods of time while trying to find jobs.

The majority of people earning US$2–5 per day are self-employed and have a limited, fluctuating income from small businesses (e.g. selling items in a kiosk, providing services such as hairdressing in customers' homes) or from taking on casual work (e.g. butcher, wire man, teacher, builder). People earning above US$5 per day tend to be employed and have a business on the side.

The major difficulty in the city, just like in low income segments in rural areas, relates to making money. Apart from food costs and school fees, rent for housing and kiosk (stall) must be paid, together with electricity and water bills.

The major difficulty is the daily struggle to make ends meet and to manage to buy staple goods (sugar, flour, cooking fat, paraffin), pay school fees (Sh 18 000–13 000 per year for secondary school) and water bills (Sh 200 per month). Monetary aspects play a part in every decision made during an average day, and the question of earning an extra Sh 30 (50c) or not could also be the difference between eating or not.

Living on more than US$2 Per Day

Syprus and Pascal, together with their two children, rent one room with a kitchenette (without running water) in a terraced house in the suburbs of Nairobi. They have their own electricity although they share water and bathroom facilities with four other households. Syprus and Pascal run their own business where they sell fabrics, clothes and shoes. Pascal also takes on casual jobs as a matatu driver or builder in order to earn extra money.

In the morning, Syprus and Pascal wake up around 5am. They get dressed and have something to eat before the children (1.5 and 6 years old) wake up at around 6am. When the children have been dressed and fed, Syprus waves off her six-year-old daughter who leaves to catch the school bus. On this particular morning, Pascal's mobile phone rings. It is Fred, who runs a matatu business. Fred needs someone to take a shift that day, so Pascal heads off to work. When he has left, Syprus takes their 1.5-year-old son on her back and walks down to their stall on the market, which is close by. When she reaches the stall she arranges her goods and chats to her friends, who also work on the market. After a little while, a customer comes along and asks Syprus if she has any women's shoes in green leather. The customer explains that she has tried to find green shoes everywhere, as she needs them to go with a new green dress that she has made for herself. Syprus doesn't have any green shoes in stock, but knows where she might be able to get hold of a pair. She tells the customer that she can probably find a pair and that she will call the customer on her mobile phone when the pair of green shoes arrives. The customer is very pleased and continues about her daily business.

Syprus makes a call to her most reliable supplier and asks about the green shoes. The supplier has them in stock and sends them over to the stall during the course of the day. When the shoes arrive, Syprus calls the customer, who says she will drop by after work to collect her new shoes. At 4 pm, Syprus' daughter comes home from school and helps her mother on the stall before going off to play with some friends.

When Syprus closes the stall for the day, she checks through her stock and realises that she needs to order some more fabric and clothes. She uses her mobile phone to call two suppliers. As she uses Safaricom's Sema tariff, she has the benefit of off-peak rates between 6 pm and 6 am. She therefore makes most of her calls to suppliers after 6 pm. When the goods have been ordered she walks home, together with the children, stopping off at the general store to buy some maize flour and milk. It is getting dark and the risk of assault and robbery is much higher now than in the daytime. Back at home, Syprus starts cooking dinner. Pascal arrives home and they all eat together. Then Syprus puts the children to bed and, after another couple of hours, the whole family is sleeping.

KEY FACTORS AFFECTING THE SPREAD OF MOBILE PHONES

The key factors that are affecting the spread of mobile communication are common among all developing nations surveyed and these are:

- Economic factors such as income per capita and the price of handsets and calls
- Flexible use of appropriate price models – for example, smaller value prepay top-up cards help overcome credit barriers and the use of mobiles as public telephones
- Government policy – mobile phone use is higher in countries with liberalised telecommunications markets
- Social and cultural factors including urbanisation, women's empowerment and population density (which can affect the cost of deployment in rural areas)

It was an idea born in those far-off days of the Internet bubble: the worry that as people in the rich world embraced new computing and communications technologies, people in the poor world would be left stranded on the wrong side of a digital divide.

'Henceforth encouraging the spread of mobile phones is the most sensible and effective response to the digital divide' (Ericsson SIM Research 2007)

"If communication is the basic human need, why do 3.5 billion people still not have access to this basic need?"

The Hard Reality

This is highly unlikely, because the digital divide is not a problem in itself, but a symptom of deeper, more important divides: of income, development and literacy. Fewer people in poor countries than in rich ones own computers and have access to the Internet simply because they are too poor, are illiterate, or have other more pressing concerns, such as food, health care and security. So, even if it were possible to wave a magic wand and cause a computer to appear in every household on earth, it would not achieve very much: a computer is not useful if you cannot read and have no electricity to power your computer.

Yet, progress is being made through the construction of specific local infrastructure projects such as rural telecentres. How the fund will be financed and managed is the subject of much discussion. An example is the Bio-fuel project where Ericsson joined hands with GSM Association and operators in Nigeria and India, ensuring that mobile connectivity can be achieved even before electricity. Even where there is electricity, it is often very unstable, etc.

One popular proposal is that technology firms operating in poor countries be encouraged to donate 1% of their profits to the fund, in return for which they will be able to display a Digital Solidarity logo. Anyone worried about corrupt officials creaming off money will be heartened to hear that a system of inspections has been proposed. This is the wrong way to go about addressing the inequality in access to digital technologies: it is treating the symptoms, rather than the underlying causes.

'The benefits of building rural computing centers, for example, are unclear. Rather than trying to close the divide for the sake of it, the more sensible goal is to determine how best to use technology to promote bottom-up development. And the answer to that question turns out to be remarkably clear: by promoting the spread not of PCs and the Internet, but of mobile phones'.

Plenty of evidence suggests that the mobile phone is the technology with the greatest impact on development. A number of recent researches have proven that mobile phones raise long-term growth rates, that their impact is twice as big in developing nations as in developed ones, and that an extra ten phones per 100 people in a typical developing country increases GDP growth approximately by 1%.

'Encouraging the spread of mobile phones is the most sensible and effective response to the digital divide and this is already proven reality.'

And when it comes to mobile phones, there is no need for intervention or funding from IDUs and donor community: even the world's poorest people are already rushing to embrace mobile phones, because their economic benefits are so apparent. Mobile phones do not rely on a permanent electricity supply and can be used by people who cannot read or write.

Mobile Phones are widely shared and rented out by the call, for example by the 'telephone ladies' found in Bangladeshi villages. Farmers and fishermen use mobile phones to call several markets and work out where they can get the best price for their produce. Small businesses use them to shop around for supplies. Mobile phones are used to make cashless payments in Zambia and several other African countries. Even though the number of phones per 100 people in poor countries is much lower than in the developed world, they can have a dramatic impact: reducing transaction costs, broadening trade networks and reducing the need to travel, which is of particular value for people looking for work. Little wonder that people in poor countries spend a larger proportion of their income on telecommunications than those in rich ones.

Mobiles Phones with High Bandwidth Access are Bridging the So-Called Digital Divide. Are We Nearer to the Moment of Truth?

According to the World Bank, the private sector invested US$230 billion in telecommunications infrastructure in the developing world between 1993 and 2003 and countries with well-regulated competitive markets have seen the greatest investment. Several firms, such as MTN Group, Bharti Group, BSNL, Reliance, Telefonica, China Mobile, AIS, Orascom Telecom, MTC Group etc. specialise in providing mobile access in developing countries. Handset manufacturers, meanwhile, are racing to develop cheap handsets for new markets in the developing world. Rather than trying to close the digital divide through top-down IT infrastructure projects, governments in the developing world should open their telecoms markets. Then firms and customers, on their own and even in the poorest countries, will close the divide themselves.

Mobile phones are proving to be the first PC that the majority of people in developing economies have ever touched. The rural villages that were without information access are truly getting connected. With the deployment of high bandwidth technologies like HSPA, people in the developing nations are very near to experiencing multimedia services like mobile, IPTV (Internet Protocol TV), video on demand and information on demand.

Mobile Success and Developing Economies

Mobile communications is fastest growing commodity in the world and the benefits of the technology have been reaped to its fullest by the developed world. Underdeveloped countries are using the technology to boost to their economies. As explained in chapter 1, the lower the GDP, then the higher the ratio of economic impact.

The 'Reverse Robin Hood' Phenomena

The success of mobile communications in developing countries can be attributed to the introduction of prepaid tariffs. Prepaid tariffs enabled the reduction of the mobile connection contract into one that could be afforded by those in developing countries. However, if we compare the tariff structures worldwide, we discover that the lower the prepaid denomination, the higher the mark-up per minute of use. If we analyse this, we quickly realise that the poorest of the poor who can only afford to buy the lowest denominations are paying at a premium; on the other hand, the high contract subscriber gets the best possible tariff per minute of use plus subsidies on handsets, or in many cases receives the latest models of mobile phones for free. This is what I call the *reverse Robin Hood* phenomena: the poorest of the poor funding the subsidies of the rich.

However, if we analyse this from another perspective, it proves the point that there is a premium to earn if you serve the consumers at the bottom of the pyramid. The unserved consumers – the majority of whom live in remote rural areas – represent a premium that if well-exploited can provide a future growth for telecom operators and service providers. In turn, these consumers get access to communication and information. This will help elevate their economic situation.

So Far, So Good. How Can Mobile Growth be Sustained?

We believe that the impact telecommunications networks have on economic and social development can be compared to the great contributions of other infrastructures e.g. roads, ports and railways that stimulate trade, create jobs and generate wealth. Leveraging this infrastructure through export-driven growth, the developing world, for the first time in history is now a net financier to the developed world.

The mobile industry can have the same profound effect for developing economies as other infrastructure but their governments and telecom regulators need to promote the right market structure for this to be realised.

When examining what the mobile industry has delivered to date, it is substantial. But let's not forget that the majority still don't have access to mobile phones, while the wholesale cost to connect to the rest of the world is often 100 times more than that of a typical DSL subscription in the USA for the same bandwidth. There are great success stories in the developing countries, but extending connectivity to all is still a big challenge for governments and the private sector alike.

As a starting point, it is important to look briefly at what the industry has delivered, to get a sense of the great potential that can be achieved if governments and industry work together towards a mutual goal.

Firstly, employment: mobile operators are becoming great sources of employment in developing countries.

Secondly, GDP: as a measure, one can use Waverman's[15] widely quoted findings that a 10% increase in mobile penetration can lead to a 0.59%

[15] Telecommunications infrastructure and economic development: a simultaneous approach *The American Economic Review*, 91(4) Sep 2001, 909–923.

increase in GDP in a typical developing economy. Ericsson, an industry leader in mobile infrastructure, believes this number is extremely conservative. Its recent research in Nigeria provides clear evidence for this belief. If one takes India as a proxy to sub-Saharan Africa, it has similar penetration levels and a large rural population. In India, the actual GDP contribution is over 1.5%. Mobile industry contributes a staggering 6.5% of China's GDP. So the incentives to connect the unconnected are there for all stakeholders.

Mobile operators are also the top corporate taxpayers and contribute additional funds through licence fees, spectrum fees, and number-range fees. In these regards, they contribute substantial sums to the national budget that can be put to wider use. Mobile networks are being leveraged to provide data services to schools, universities and hospitals.

Who would have thought that mobile operators are the largest ISPs in many developing countries? SMS messages are frequently used to increase the effectiveness of health programmes. Mobile customers are benefiting from access to financial services for the first time as they use their phones as virtual banks. Higher mobile penetration really does increase the governance, social capital and economic activity of nations.

So How Can We Reach the Next Three and a Half Billion Users?

Simply put, by eliminating fiscal and regulatory bottlenecks the industry's cost structure will fall and penetration and usage will increase as services become more affordable. The elasticity of demand for telecoms services is huge.

Goldman Sachs estimates that a minimum per capita income of over US$1000 is required to afford mobile services in the current paradigm.[16] Given income distribution, the investment bank concludes that we will have to settle for relatively low penetration levels in Africa. However, the GSMA is working on projects to provide mobile services to a market with sub-US$100 income, and the GSMA believes that a true public-private partnership will prove the bank wrong.

Four Main Tools

A quote from Ambassador Gross, the US Coordinator for ICT, illustrates the first policy tool. He said that *'If you want to encourage something, don't*

[16] www.developingtelecoms.com/content/view/481/100/

tax it'.[17] As with other businesses, mobile operators pay significant amounts in corporate taxes, but also pay additional fees for things such as licensing, spectrum and number ranges. Consumers are also being asked to pay specific mobile taxes on handsets and services. This renders them unaffordable for many aspiring potential customers and shrinks the addressable market for operators and also, therefore, the potential revenue for governments.

For example, the Ugandan regulator has done a fantastic job ensuring some 92% of the population has mobile coverage. Why then does the Ugandan government levy some of the highest taxes on its mobile customers so that only around 10% are able to benefit from the expensive network?

Secondly, in the recent GSMA study *Regulation and the Digital Divide,* data from 28 African countries was analysed. The study clearly demonstrated that erratic regulation increases the cost of capital substantially, which is a major proportion of the cost base of a capital-intensive industry. The study also found that if consistent and fair regulatory practice had been in place, the costs of capital would be lower and an additional US$5 billion would have been invested in mobile capacity and infrastructure. That is equivalent to the accumulative capital expenditure of both MTN and MTC Celtel, the biggest multi-country operators in sub–Saharan Africa. Imagine the increased coverage and penetration this would have brought about.

Thirdly, monopoly controls on international gateways must end. These are choking African businesses as they seek to compete in a global market place. And they substantially increase the cost of doing business regionally too.

Fourth, and of critical importance, is the urgent need to bring the cost of connectivity to the rest of the world down through open – access fibre – optic cables. These cables are the umbilical cords that Africa can rely on to grow, stimulating the creation of new industries and employment, increasing the continent's competitiveness, providing fast, reliable and affordable Internet connections for students and businesses, and encouraging FDI and export – led growth.

We are all familiar with the Indian BPO success story. It will come as little surprise that Indian companies own more than one third of the undersea cables around the globe, ensuring that India's umbilical cord is as wide as necessary for the local economy to grow. There is no reason why this type of success cannot be replicated in Africa.

---

[17] blogs.nmss.com/communications/2006/02/telecom_for_rur.html

Connectivity – Do Not Throttle Africa at Birth!

The structure of the Sat-3 cable network has kept bandwidth prices artificially high. Telkom has been charging US$25000/ Mb/month until
recently, outrageously high compared to what one pays for a DSL connection in a country like the USA – about US$30 per month. No wonder a
large portion of available capacity on the cable is unused; the West African
umbilical cord is being throttled.

There is a real threat that the Eastern Africa Submarine Cable system
cable may go the same way. However, anecdotal evidence indicates that
the East African governments do not want to repeat the mistakes of Sat-3.
The work of Kenya's Minister for Informatiion and Communication
Mutahi Kagwe and his East African supporters who are highlighting this
issue is to be applauded.

> *The cable that connects East Africa to the rest of the world must have an open access*
> *structure for the region's potential to be realised.*
>
> *Less tax = more subscribers.*

Returning now to the role of taxation, it is evident that there is a close
correlation between mobiles taxes and subscriber growth. Our GSMA tax
study also found that:

- The mobile industry pays a higher share of tax than the fixed;
- About a third of handsets are sold on the black market in Africa as
  users try to avoid paying tax; if low-cost handsets were exempted
  from import duties and sales taxes an incremental 930 million handsets would be sold over a five-year period;
- Lower taxes mean greater long-term revenue opportunities for governments; and
- Cutting taxes on handsets would attract new users who would each
  yield an additional US$25 in annual tax revenue.

Lowering taxes is not the only method that can help foster growth; regulation also plays a role in the development of the mobile industry. The
GSMA regulation study showed that a regulatory environment that
treated mobile operators fairly and consistently, reducing uncertainty and
enabling a longer – term investment horizon would:

- Increase sector investment by 25%, i.e., US$5 billion – the sum of MTN
  and Celtel's CAPEX;
- Increase penetration by 30%; and
- Boost regional GDP by US$1 billion.

The 'Digital Dilemma' for Governments and Regulators

High taxes on mobile services run counter to governments' commitment to improving access to communications. At the World Summit on the Information Society in 2003, 175 countries signed up to a commitment to give more than half the world's population access to information and communications technologies by 2015. If governments took the right approach to taxation, that goal could be achieved within five years, yielding huge benefits for developing countries and their people.

Innovative Approaches to Mobile Taxation

Earlier this year, the GSM Association (GSMA) set in motion a programme to help eradicate the barriers to accessing mobile communications for people in developing countries This programme was the direct result of research that identified the cost of the mobile phone as one of several barriers to affordability in these nations.

The GSMA challenged the manufacturing sector to respond and deliver to the market an *ultra low cost* mobile phone. Initially this was achieved at a sub-US$40 cost, although through further innovation, the cost has since been reduced to sub-US$30 and it is going further down to the range of US$15–20. Thus, a new low cost market segment, previously unaddressed by the industry, is being created. Operators are offering innovative pricing models that make it even easier for consumers earning below US$2 a day to afford it.

However, it has been clear that taxation would form a key additional component in the total cost structure of mobile products and services. To understand the full extent to which taxation has a direct impact on affordability, the GSMA commissioned an independent study across 50 developing countries.

The results are surprising in terms of the degree to which taxation acts as a barrier for users, preventing potentially hundreds of millions of people from affording mobile communications, and holding back economic growth and social development in many countries.

The study's key findings are:

- **Taxes are disproportionately high in many developing countries**
  - In 16 of the 50 developing countries in the study, taxes on mobile phones and services represent more than 20% of the total cost of ownership. In these 16 countries, which are home to hundreds of millions of people, the annual cost of taxes ranges from an average of US$24 to US$179 per mobile phone user.

- Nineteen countries even levy additional taxes, on top of standard sales taxes, on mobile phone users. Some of these additional taxes are telecom specific, such as service activation taxes. These special taxes average US$13 per annum per subscriber.
- **The black market in handsets is booming as users try to avoid high taxes**
  - 39% of all handsets sold in the 50 countries in the study in 2004 were via the black market, representing a loss of $2.7 billion in tax revenues.
- **Cutting taxes on mobile handsets and services attracts new users**
  - If low cost handsets were made exempt from import duties and sales taxes, up to 930 million additional low-cost handsets could be sold by 2010 in the 50 countries in the study, leading to an increase in mobile phone penetration and a rise in total tax revenues in some countries.
  - If a government lowered taxes on mobile usage by just 1%, that could boost the number of mobile users in that country by more than 2% by 2010.
  - Eliminating the special taxes could boost the numbers of mobile users in the 19 affected countries by 34 million (or 8%) by 2010.
  - The removal of all sales and customs taxes on mobile handsets and services could prompt an increase in mobile penetration of up to 30 percentage points, according to an internal analysis of sub-Saharan Africa by Ericsson.
- **Lower taxes mean greater revenue opportunities for governments in the long term**
  - Cutting taxes on handsets would attract new mobile users. If taxes on usage remained the same, each new user could yield additional service tax revenues of US$25 per year.

Kick-Starting the Developing World's Drive for Growth

Everyone has an interest in eliminating poverty in developing countries, and that means tackling long term issues as well as short term problems. A key challenge is to improve the economies of developing countries – especially the lives of the less wealthy members of their societies.

The United Nations has set the tough challenge of halving global poverty by 2015. Harnessing the potential of Information and Communications Technology (ICT) is an integral part of the United Nations' Millennium Development Goals, which has led to the creation of a Global Alliance for ICT and Development. This is designed to provide a platform for public and private partnerships, as well as a network of international experiences.

Many studies show how important ICT is to development and growth. In this context, mobile phone communications is one area that can make an immediate and direct impact both on the daily lives of the poor and, just as important, on the business economies of some of the world's most needy nations.

Universal Access

When reaching the poor with ICT capabilities, there is a difference between providing *Universal Access (UA)* and delivering a *Universal Service (US)*.

Introducing UA means giving people reasonable means of access to a publicly available telephone, SMS and emergency service in their communities – but not necessarily in their homes – through shared use of terminals, including public payphones, community centres and cyber cafés. Universal access is a strategic policy for low-income and/or high cost areas where private demand and perceived need do not always attract service providers to invest.

Nevertheless, today in the vast majority of countries a high percentage (often more than 95%) of the total population can be reached with GSM – a feat not achieved with fixed communications.

Universal Service

Universal Service is concerned with providing basic telephony to every household, thus overcoming the social and economic disadvantages associated with not having private access.

With mobile service penetration already at between 50%–75% of households in many developing countries' urban areas, the achievement of universal service is a realistic short-to-medium term target. Many operators already offer low/affordable tariff packages so the main hurdle to overcome is often the price of the handset.

GSM, the leading mobile technology with its natural evolution path to 3G and high speed data capabilities, has grown to become the most popular end user technology in the world. With over two billion subscribers and almost one billion phones produced this year, the economies of scale have brought phone prices down to US$30. At the same time, new mobile network features have reduced the total cost of ownership for the service providers. This evolution is making mobile services affordable to the low-spending segment of the market.

The rapid transition of GSM networks to EDGE (Enhanced Data Rates for GSM Evolution), and also 3G, means mobile operators can also offer the capability of enhanced data, fax, Internet and other ICT services.

It is the collective strength of providing universal access while offering affordable universal service to institutions and individuals that characterises the benefits of wireless systems.

The Challenges

Mobile networks are being deployed in developing markets – in both urban and, increasingly, rural areas. The challenge is to create the right regulatory, taxation and competitive environment and put the right technology solutions in place.

To achieve high penetration in developing markets, person-to-person communication has some big advantages over other forms of ICT. Users do not need to own a PC, or have access to a fixed line infrastructure. Nor do they need to be computer literate or have particular language skills.

A study in rural Tanzania showed that 97% of village populations knew about mobile telephony, with half of them having already used a mobile phone. By contrast, some 67% did not know what a computer was and only 3% had ever used one (source: UNDP & Swedish Agency for International Development Cooperation).

With access to affordable handsets, shared usage and low tariffs, mobile communications can be made widely available, especially as many people use their phones primarily to receive incoming calls. Low top-up denominations for pre-pay services appeal to consumers with less than a dollar to spend. At the same time, it must also be simple for them to sign up and stay connected.

Once mobile systems are in place, SMS messaging is available and can account for a significant part of operator revenues. It is not unusual for the SMS volume to significantly outnumber voice calls in rural communities.

Internet access can also be provided for isolated schools, hospitals, police stations and farms, for example. PC cards in laptops can provide data and fax services via the mobile network for more advanced users. Initially, Internet access can be delivered by GSM/EDGE technology but this can later be upgraded to WCDMA/HSPA (wideband code division multiple aceess high speed packet access).

Of course, the only sustainable business model for delivering mobile communications services is one that is profitable for the service providers concerned. These days, there is a range of technical solutions available, as well as the opportunity to share network facilities to serve rural and remote areas in developing markets. In this context, new business models – where one of the operators, or an independent operator, manages and delivers the required network coverage and capacity to the whole market – are emerging.

Lack of power and transmission networks can be a challenge in rural areas but these problems can be overcome by renewable energy sources and possibly satellite links, eventually replaced with lower cost high speed microwave links.

Governments' tax and regulatory regimes are also critical. The temptation to tax mobile phones and services as luxury items needs to be resisted. A GSM Association survey indicates that if all taxes and duties were removed from mobile handsets, penetration would rise by 20%. The commercial environment is also vital: open competition, with adequate safeguards, is needed to ensure low prices, choice and growth.

An African Example

Mobile operators around the developing world are expanding their networks to handle more subscribers, more cost-effectively. For instance, sub-Saharan Africa, which seven years back (except for South Africa) was almost unconnected, has been the world's fastest growing region for mobile communication over the last seven years and growth still continues.

Several African mobile operators are now moving from high end ARPU focus to low ARPU at the lower end of the consumer pyramid and hence will drive growth in the rural remote areas. Improvements are being made to networks in several countries, including Sudan, the Democratic Republic of Congo, Sierra Leone, Gabon and Chad. Mobile operators are upgrading their GSM networks and bringing signals to areas which have never seen any kind of communication infrastructure. Using high-capacity base stations, operators have been able to reduce the number of sites needed for initial roll-out and thereby achieve flexible capacity and coverage expansion, resulting in better coverage for lower investment.

The possibilities for mobile communication-a critical first step in applying ICT in developing countries – are significant. This should be reassuring to all concerned, and an important focus for the United Nations Global Alliance for ICT and Development (UNGAID) global initiative.

THE PHENOMENAL GROWTH OF MOBILE TELECOMMUNICATIONS – WHAT IS HINDERING FURTHER DIFFUSION?

The following are the key obstacles to realising the ambition of sustaining the ongoing growth in mobile communications and achieving the goal of providing access to mobile phones for all.

- **The cost of mobile phones**

The cost of mobile phones has been declining substantially over the years. A number of initiatives are being used by the industry to drive the handset cost down. Today, mobile phones are available in the range of US$15–20 which must go down further to a US$10 level. Service providers must join hands with the banking industry to find a way of financing the handsets, breaking down the initial payment into small monthly installments.

- **Mobile tariffs**

Mobile tariffs have been going down significantly over last few years, but in order to reach this new wave of consumers the tariffs need to go down further. Service providers need to lower the cost of production, which can be done by using number of different cost reduction techniques like network sharing, using efficient processes, outsourcing non-core functions of the business and operation. Low Tariffs / ARPU do not necessarily mean low margins; some of the best providers in emerging markets have superior margins with ARPUs3x–5x lower than in more mature environments. Figure 2.2, below clearly demonstrates that the low ARPU does not mean lower profitability. They tweak their business models to exploit the bottom of the consumer pyramid.

- **High prepaid denominations**

It has been proven that lower the prepaid denomination the higher has been the usage. Service providers must provide the lowest possible prepaid denominations, in fact they must offer flexibility of buying the top-up value of any denomination.

- **Service provider business and operations models**

Service providers need to transform their business and operations models to ensure the lowest cost of production, higher grade of service and network quality, advanced service and products customised to the needs and aspirations of the consumers.

- **Lack of network coverage and capacity**

Lack of network coverage and capacity is another obvious barrier to diffusion. There are many people living in areas without coverage who are

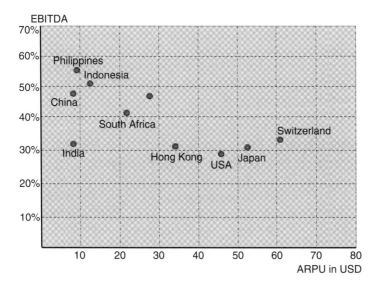

Figure 2.2   Low ARPU does not seem to reduce profitability

very aware of the income-generating and time-saving benefits that come with network coverage and mobile phone ownership. Many of these people say that they would save up the money to buy a mobile phone if coverage was available in the village where they live.

Network and technology infrastructure providers need to invest in research and development to find solutions that suit this new wave of users, create robust technology that is less dependent on power and air-conditioning and to create technology that boosts the network coverage and reduces CAPEX per subscriber.

• **Expiry dates a hindrance**

Airtime expiry dates present another barrier that can prevent low income households from buying a mobile phone. As the top-up card expires just after 10 days, there is little incentive for people who make very few calls to save up the money and buy airtime. It is not uncommon for low income house holds in villages to make only 1–2 short calls per week. When these calls last less than a minute and are made at off-peak rates (say Sh10 per minute), a Sh50 scratch card expires before the user has even consumed the total airtime.

Before we tackle the above barriers, let us try to understand who and where this growth will come from. Majority of the new wave of users come from the bottom of the user pyramid, people with income less than US$2 a day. Many of them cannot read and write. Majority of them live in remote areas that lack supporting infrastructures like roads and electricity. Lots of them don't have access to cash economy.

The million dollar question is: how do telecommunications companies find a business case to justify the investment in bringing communications to these consumers. The answer lies in the following: the profitability models need to be reexamined.

Adjusting Service Provider Business Models for Serving Consumers at the Bottom of the Pyramid

The traditional profitability model does not apply when delivering the business case for serving consumers at the BOP. The traditional model for deriving profitability is based on the simple equation: Profit = Revenue (ARPU* Number of Subscribers) − Cost (OPEX + CAPEX). This model takes into consideration all cost upfront to drive the business case and is shown in figure 2.3, below.

However when approaching this new wave of users we need have a different mind set. The business / profitability models that took the global mobile subscriber base to 2.5bn subscribers are distinctly different from those needed to take that same user base to 5bn and more. One needs to apply the model described in figure 2.4, below:

CAPEX and ARPU are not part of the equation. The profitability is driven by increase in volume and efficiency in OPEX. We call this the *volume game.*

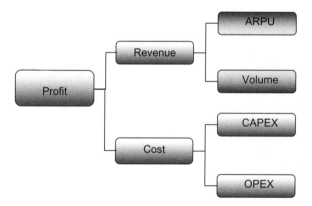

Figure 2.3 Traditional model for profitability

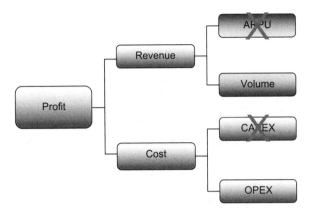

Figure 2.4   Adjustment to the traditional model

The Volume Game

When approaching consumers at the bottom of the pyramid and setting an ambition to reach this mass market, we as an industry have to change the way we have traditionally approached the business. We have to rethink our business models ensuring that we are removing all the obstacles to uptake. Removing frills and offering a basic service to ensure lower prices are not the solution. People living in these far flung areas do have needs for high end services; the only problem is that they can't afford to pay the tariffs that are applied in the developed world.

We as providers need to develop business models that are driven by economies of scale. This is what I call the volume game. We must lower the tariffs for basic services to a level that are affordable. We then offer value added services and take premium there. Apart from that, we need to evaluate and embrace the operating models that help minimise the OPEX and CAPEX spend.

Role of Governments and Regulators

Mobile phones have the potential to give billions of people in developing countries access to communication and information technology, but only if governments work with the mobile industry to reduce the total cost of owning and using a mobile phone.

For example, the Indian government has already demonstrated this isn't just a theoretical notion. It has brought down handset import duty over the past five years, helping to boost mobile penetration from less than 5% to more than 15%. However, other fiscal barriers to mobile usage exist in India.

Previous studies (Ericsson SIM Research 2007) have shown that mobile phones play a major role in reducing the cost of doing business and driving entrepreneurialism across the many developing countries without a widespread fixed-line infrastructure. Mobile phones make it much easier for farmers, fishermen and a host of other business people to find buyers and sellers for their products and services. In fact, an increase of 10% in mobile penetration boosts a country's annual economic growth rate by 1%, the impact is higher, the lower the GDP of the country. For example, the relations were found to be 3 fold higher in Nigeria and some other African nations (Democratic Republic of Congo, Angola, Bangladesh, Pakistan).

The mobile industry has made considerable strides in driving costs down through lower handset costs and innovative service solutions for lower income groups, as well as extending mobile coverage to 80 to 85% of the world's population. More still needs to be done. In the light of the findings of this study, more governments now need to take up the baton and re-evaluate the impact of their tax policies on mobile communications. Governments and mobile operators should work together to determine the ideal tax levels for their particular countries.

Transforming Cost Models

Owing to the demand constraints impacting revenue, low ARPU business models are as much about innovative cost management as they are about driving the revenue growth. Telco's can directly influence two major cost lines: selling and general administration, and network operations and maintenance. Together these two lines account for 40 to 50% of the operator OPEX, the other key cost elements are the interconnection fee, any applicable license fee, handsets and accessories, which can be significant if the market requires heavy subsidies.

Case Study 3: *India Paving a Road for the Mass Market*

Introduction

With the lowest tariffs in the world, India is the master of low-cost operations. Low tariffs are generating mass market appeal. In fact, India is experiencing the fastest mobile subscriber growth, soon reaching 200 million subscribers, growing at a rate of over 7 million subscribers a month.

The Role of the Regulator

The Indian regulator is working with all players in the mobile market, operators, suppliers and authorities with the objective of establishing and sustaining a good mobile infrastructure and enabling affordable mobile communications for the mass market in India.[18] Recent reports indicate that the Indian government aims to expand teledensity to more than 500 million users by the end of 2010 – around one in two of the population.

The timing is good. The vast majority of India's people live in rural areas and, at the beginning of 2007, only about 60% of the population was covered by a mobile network, a fact acknowledged by the government which has declared that expanding population coverage further is a priority.

The government and changing regulatory environment is catalysing the faster adoption of wireless technologies in India. In fact, the Indian mobile market is approaching a sharp upturn as favourable demographics, strong economic growth and cheaper handsets position the country to deliver strong results in the next several years, although again, that may depend not just on how the market evolves but whether the regulatory environment evolves further to facilitate it.

To reach the ambition of 50% mobile penetration in next three years regulator has introduced following initiatives:

Universal Service Obligation

Under this initiative, 5% of all revenues from operator and license fee are made available to private operators that want to expand their rural networks and coverage. The fund currently is estimated to be over US$1bn.

Network Sharing

The government and regulator have introduced a system of mandatory sharing of base stations towers. There are over 8000 sites around the country, all subject to an individual tendering process with competing companies. Every new rural tower has to be shared by three operators. This cuts CAPEX dramatically and reduces OPEX for running these sites and in the end reduces the total cost of ownership for all the operators. Also to limit numbers of towers in the urban areas 6 to 7 GSM and CDMA operators share each tower; achieving the same objective as in the rural area lowest TCO.

---

[18] Jansson, H. (2007). How India reached critical mass for mobile services. *Ericsson Bussiness Review*.

Plants Are Used to Fuel Networks

One of the clear obstacles when building rural coverage is the lack of supporting infrastructures, for example, electricity. In many rural areas, mobile connectivity is being achieved before electricity infrastructure. Even where there is electricity it is often very unstable. To counteract this problem, operators such as IDEA in India together with GSM Association and Ericsson have started a Bio-fuel project. This project ensures that, by using plants, electricity is generated locally in rural areas, avoiding the need for transportation. If this is done at a commercial scale, it will allow access to the market, where otherwise this would not make economic sense.

From Voice to Enriched Communications

With globalisation of data networks such as GPRS, EDGE etc. a number of innovative data services projects are being launched to meet the desires and aspirations of the mass market. Joint projects are being initiated with the banking industry to enable m-commerce applications. This is revolutionising the banking industry, enabling access to modern banking for millions of people.

Tariffs and Operator Economics

Despite tariff levels as low as US$0.01 per minute, Indian mobile operators in general are coping and businesses are healthy. Despite low ARPUs an average blended ARPU of US$7–8 and a marginal ARPU of US$3–4, operators are showing healthy earning before tax levels. The question is: how do they manage to do this? After talking to some c-Level executives in India, it was apparent that they all manage to bring down the cost of production by embracing innovative business models. They have managed to eliminate inefficiencies in their operations simply by outsourcing all non-core elements of their operations such as IT, network etc.

Provide Empowering Services

This enables profitable micro-businesses. People in lower-income segments are willing to spend a higher portion of their income on mobile communication because it opens up new income opportunities, especially for the self-employed and micro-businesses. The principle is 'you have to spend money to make money'. Empowering applications include the ability to access information, make financial transactions and to pay with transferred airtime.

Mobile Technology Evolution, Relevance for Next Three Billion Users

Operators who deploy High-Speed Packet Access (HSPA) technology have several advantages over those who choose different paths for delivering advanced services. Today 130 operators in 60 countries have already committed to this technology, proving HSPA is the obvious and quickest route bringing mobile broadband to the mass market. By a software upgrade, a WCDMA network can be enhanced with HSPA. With 85 million people around the world already subscribing to WCDMA services, no other high-speed radio access technology has a comparable installed base.

The same advantages that made GSM the catalyst for mobile telephony achieving the mass market also apply to HSPA. The combination of legacy GSM networks, economies of scale and a vast selection of consumer devices ensure that people will have access to multimedia services, the Internet, TV and games on any screen and device of their choice.

To date, the amazing growth of mobile communications has come primarily through voice traffic. As the barriers to mobile data usage such as cost, speed and quality of content are removed, operators will need to react to more demanding, dynamic users and their requirements.

Operators must take a holistic approach so they can provide any broadband service to any device a subscriber wants no matter where the user may be. Users have become accustomed to dependable, easy-to-use services on the Internet, and mobile operators must work to ensure the same quality in a mobile environment.

This creates, in effect, a new channel for distributing content. Mobile TV is already available on networks around the globe, and music, gaming and online communities that are successful over the Internet are likely to follow. As the number of mobile subscribers in the world is approaching 3 billion, we can see the mobile content phenomenon is still at a very early stage.

Equally important as the high data speeds possible with HSPA networks is the service integrity provided by IMS. Together, these new technologies will guarantee the quality of service (QoS) users expect, allowing broadband services to be available anywhere on any device in a quick, cost-effective fashion with minimal risk. This ensures the integrity of multimedia services across all access methods.

CONCLUSION

Despite of enormous growth in mobile communication, more than half of the world's population still do not have access to basic mobile communications. The majority of this new wave of mobile users, people at the so-called bottom of the economic pyramid, have income levels less than US$2 a day. Hence, affording the luxury access to mobile communications, although a basic human need, is beyond their means. Owning a mobile phone and affording the mobile services is still an aspiration for many people. Most of these are very poor and the majority of them live in rural regions. The benefits of the digital age have not filtered down to the majority of people in the underdeveloped economies.

In view of the economic and social benefits of the development of the mobile sector it is important to identify the current obstacles to development. The problem is not a lack of demand or willingness to pay. Handset prices are falling and innovative mechanisms have been devised to bring much-needed services (communication and otherwise) to even the poorest individuals. Nor is it a case of technological obstacles. Technological solutions exist to cover countries of any size, shape and terrain with high-quality communications. The cost of deploying and launching basic services is constantly decreasing due to economies of scale in technologies, in particular GSM. This evolution brings increased affordability in urban, as well as rural areas.

The main barriers to development result from a combination of several factors. These include significant investment requirements, considerable regulatory and other risk associated with these investments, low income levels, and areas for which service provision is extremely costly.

In the short to medium term, some of these factors such as geographic and income-related issues will remain problematic. What can be addressed is the matter of regulation and the associated risk. A sustained effort by regulators and governments is needed to provide a climate of stability in which the communications sector can thrive further and henceforth help realise the dream of communications for all. The main role of governments will be the establishment of an appropriate regulatory regime, the provision of essential infrastructure (roads and electricity), and the abolition of short term penalty or luxury taxes levied on the mobile sector. The main role of regulators will be to implement best-practice regulations that are appropriate for, and tailored to, the countries concerned, and to do so with a maximum degree of predictability.

Increasing capacity and coverage to reach these users must be based on profitable business principles. Using a traditional business approach to address this challenge will never make economic sense. Service providers and the supporting industry need to master low cost operations, innovate technology that reduces dependency on supporting infrastructure e.g. Bio-fuel, solar panels for energising mobile base stations, low cost handsets etc., create products and applications that are relevant and compelling for this new wave of users.

All stakeholder governments, regulators, telecom service providers, technology providers, financial institutions and donor communities need to be invoved: governments must reduce the tax burden, encourage network sharing, provide incentives to got to the far flung areas; operators must create efficiencies, use techniques to master low cost operations; technology providers must come up with innovative solutions that reduce total cost to ownership; financial institutions must create innovative funding models; and last but not least, the donor communities must move away from theoretical contribution to practical contribution e.g. Ericsson, World Bank and SIDA joint initiative in Tanzania.

In conclusion, what we can achieve is the eventual removal of the digital divide by mobile industry growth and, as the GSM technology evolution-path plays out, this shall bring not only voice and affordable data communications to hundreds of millions of the world's poorer people, but also enriched multimedia services like mobile Internet, instant voice and video messaging, mobile TV and whole host of value added applications for everyone. To achieve this goal, affordability must increase and the total cost of both ownership and use of mobile services must fall, tapping the great latent demand.

3

Communications for All – Is It a Myth?

WHO ARE THE KEY STAKEHOLDERS?

Examining the developing countries, you will find that more than 50% of the world's population doesn't have access to either voice or data telecommunication. The majority of these people are poor and live in rural areas without infrastructures such as banking, electricity, roads and communications. A large number of studies have been performed by the Swedish Agency for International Development Cooperation (SIDA), UNDP, operators and other institutions.[1] The studies all point in one direction: the so-called *killer application* for poor people in rural areas is mobile voice communication. It is also clear from these studies that mobile voice communication is a prerequisite for sustainable economic growth in these areas.

We also know that extending current mobile networks to the rural areas is not happening fast enough to satisfy the ICT objectives in the UN Millennium Goals. However, there are good reasons why operators don't extend networks to rural areas. First, they earn good money where they are today, in the cities, and the current business models used today confine wireless solutions to high density areas. Secondly, many operators have OPEX issues and if they would enter the rural market, they have to build their own networks and operations in parallel with their competitors. Therefore, they are not so eager to take the financial and operational risks of entering the low ARPU (Average Revenue Per User) regions.

[1] For example: Mendes, S., Alampay, E., Soriano, E., and Soriano, C. (2007). The innovative use of mobile applications in the Philippines–lessons for Africa. SIDA Edita Communication. Available from www.sida.se/publications

Business Models for Sustainable Telecoms Growth in Developing Economies
S. Kaul, F. Ali, S. Janakiram and B. Wattenström
© 2008 S. Kaul, F. Ali, S. Janakiram and B. Wattenström

The idea of communication for all will become a myth unless stakeholders get together and start cooperating toward a common goal. Especially when we address rural areas in developing countries the criteria are very different to the developed world, so the normal rules and models don't apply.

However, it is important to view this market from a commercial angle for the sake of sustainability. As in all business, there will always be an element of risk involved, but the risks can be shared with more stakeholders. Creating business models that spread the business risks will attract new investors to fund new ideas and initiatives to provide communication for all. In order for these initiatives to be successful it is important that stakeholders apply an outside-in view. This means that the business model must create value for the consumers and provide an opportunity to increase their quality of life.

The stakeholders that need to work together are:

The Politicians

The politicians need to show a clear ambition to provide the rural poor with basic communication. The political decision makers can play a key role. Barriers such as taxes and policies that prevent communications networks being built, especially in the poorer areas, need to be removed. A tax on communication is a tax on growth. There are two very strategically important areas that should be on top of government agendas: liberalisation of the telecoms market and the independence of the regulatory body. In countries where the telecoms market has been truly liberalised and the regulatory authority has been able to do its work independently, the market has taken off, bringing substantial GDP growth to those countries.

Telecom Operators

The telecom operators are key to the establishment of rural communication. The operators need to be open to new ways of doing business, involving sharing financial and operational business risks with others. Business innovation is becoming an increasingly important focus area.

Telecom Vendors

The vendors will provide new innovative business models and technologies to lower the total cost of ownership. It is important that the solutions provided by vendors are tailored to meet the specific criteria in developing countries and rural areas. To drive the total cost of ownership down further, it is necessary that vendors broaden their scope and incorporate strategic and innovative partnerships across business segments.

Telecom Regulators

The regulatory authority will be key as they drive the regulation in a country. They need to level the ground in providing licenses, spectrum and land, and take a leading position in facilitating and supporting the implementation of new business models. Also rules and regulations in some developing countries are unclear and possess a barrier for growth.

A regulator has significant power to affect the level of risk faced by the operator. Regulation matters to every aspect of a mobile business, so investment decisions will be closely tied to the cost and uncertainty associated with it. Regulatory powers tend to include the regulation of interconnection tariffs and the imposition of fees for licenses and additional spectra. A significant amount of uncertainty is created where regulators either renegotiate previous agreements or include large fluctuations in key parameters (for example, by rapidly changing interconnection tariff levels). The reputation of a regulator is crucial to establishing a low-risk investment climate for operators. All aspects of regulatory policy will influence willingness to invest including:

- Interconnection rates
  Whilst it is important to ensure that the right rates are set, it is also important to avoid disproportionate fluctuations in interconnection fees. Where interconnection rates are set based on benchmarks, it should first be ensured that the countries used in benchmarking are comparable in terms of geographic, economic and technological factors. Similarly, where cost models are used, it is important that the input parameters are appropriate. If these safety measures are neglected, then interconnection rates may end up being *corrected* up and down in an unpredictable manner.
- License or spectrum withdrawal
  In some countries, governments have threatened to withdraw licenses from existing holders, or have revoked spectrum that has already been allocated. For example, MTN and Atlantique Telecom in Benin had their licences withdrawn in 2007, and the regulator said that they needed new licenses with much higher license fees.[2] In some cases, this has led to mobile operators putting all investment projects on hold.

[2] See www.iweek.co.ea/ViewStory.asp?StoryID=176841

- Imposition/variation of taxes and fees
  In some countries, unexpected fees have been introduced to charge
  numbering ranges. In other countries, tax levels which were agreed
  in advance have been increased significantly after the entrant had set
  up its operation. In one instance, the mobile operator wrote off all the
  investments undertaken and ceased operation in the country.

NGOs

The NGOs need to take an active and positive role in supporting new
initiatives and be involved in providing expertise and funding when
needed during a start-up phase. A closer partnership between the private
and the public sectors will be crucial for bridging the digital divide.
However, this is not the case today. The private and the public sectors still
seem to be too far apart. The NGOs need to understand that the private
companies could bring substantial socio-economical values to the local
communities in the emerging markets. It is also important that the NGOs
understand that commercial telecommunication investments in poor
areas of these countries will also empower local entrepreneurs and companies, bringing new opportunities and jobs to theses communities. Like-
wise, the private companies need to learn how to establish partnerships
with NGOs and how to use the NGO local competence for the benefit of
these projects.'

World Bank/UNDP/UNEP/World Economic Forum[3]

These organisations are examples of organisations that could take a
leading role as neutral facilitators. Distrust and doubts between the commercial and the non-commercial sectors hinder good initiatives. These
organisations could and should act as a bridge to form working partnerships between the sectors. We need to remove red tape and build trust for
a common cause that will benefit both these organisations, private enterprises (both local and global) and consumers. Indeed, there are now signs
that there is an increasing understanding of the need to work in new
partnerships especially to bridge the digital divide.

Investors

Both local and international investors are needed for the financial set-up.
Experience shows that with the right stakeholders involved and the right

[3] United Nations Development Programme (UNDP); United Nations Envirion-
ment Programme (UNEP).

business models/cases, there is no lack of willing investors. The name of the game is risk management. A significant part of new business models for bringing sustainable rural communication to developing countries is the financial model. All solutions needs to be financed. The problem is not lack of money, but rather lack of trust and sound business cases. Another important condition for attracting investment is the political and regulatory stability of a country.

Management and other decision makers need to *think out of the box* and see the challenge of providing the rural poor with mobile communication from a new angle. Traditional roles may need to change for the good of development. It is important that all aspects are covered in the process, including ensuring participation by local communities.

One maxim that I found especially useful for the thought process is: *if you want success, watch what the masses are doing and do the opposite.* Most people are not successful. The successful people are those who dare to find solutions in new ways, trying a different angle.

Local Communities

When applying new business models, especially in rural areas within a developing country, it is critically important to involve local communities. Clearly, they are one of main stakeholders in securing a sustainable communications solution. Quite often, operators find that fibre optic and copper cables as well as diesel from base station generators will be stolen by vagrants. As a direct result of these kinds of happenings it can become very costly for an operator to set up a mobile infrastructure in these areas. However, these obstacles can be overcome by placing guards at the base station site on a 24hour/7day basis, and a number of other security measures installed simultaneously that make the business case difficult.

One way to better protect the infrastructure at the same time as we lower the total cost of ownership is to involve the local communities. Examples of this could be to involve local communities in the maintenance of the access infrastructure. This also leads to a need for technical training that, in turn, will contribute to educating local people and the provision of jobs. Another way of involving local communities is the production of biofuels. Locally produced biofuels can then be sold to the infrastructure owners.

This creates local jobs, creates a local economy and increases the educational level in local communities. This will, in turn, lead to the local communities protecting the infrastructure, as it represents a major source of income for them.

HOW CAN THE DONOR COMMUNITY HELP?

When addressing rural areas in developing countries, the telecom vendors do not have detailed insights into the daily living conditions of poorer people in those countries. The ability to use communication for emergency and family networking as well as getting access to information for financial transactions, substitute traveling, improved education and health and access to Internet will dramatically improve living conditions for financially constrained people. This we have learnt through studies.

Many studies have been performed, especially for the sub-Saharan part of Africa, and we now have enough knowledge to act. Most of the telecom related activities from the donors today are focused on more studies or introducing new Internet based projects. These actions will not help the rural poor nor will it help toward the UN Millennium Goals. The target date for the UN Millennium Goals for Africa is set to 2015, but with the current pace we are looking at 2147!

What we need to do is to work together, NGOs and the private sector. Projects totally paid for with donor money will not work, but we need the local expertise from the donor community and we sometimes need donor financial support to start new initiatives (seed money). Otherwise these projects may never start. Whenever you explore new grounds, financial support is needed.

However, there is a historical gap of trust between the public and private players that prevents a working partnership. NGOs often view commercial companies with a great amount of suspicion. They often lack operational decision making power especially if there is a commercial component involved. The commercial vendors on the other side don't see the value that the NGOs can add to a commercial project and therefore don't make any efforts to engage the NGOs.

The donor community can play a significant role in providing communication to the underserved in developing countries. It can be done in a sustainable way if they dare to work with the private side. In other words, there is a lot to gain for all and especially for poor people in the developing countries.

The donor community can:

- Provide local expertise to better understand local consumer needs in underdeveloped areas.
- Monitor social and economic development based on commercial projects.
- Increase local knowledge to these projects.
- Actively support start-up of commercial projects.
- Support the local involvement of application development.
- Facilitate local contacts to other local stakeholders. It is important that the local community gets involved in commercial activities. I believe

that a successful commercial project must involve local people and companies.

A first step, if this is to happen is that both the public and the private side need to make an effort to understand each other better. That effort could be facilitated by a neutral body such as the UNDP or World Economic Forum.

Importance of Profitability

Regardless of the outcome of discussions over the proper level of development assistance funding from the world's wealthier nations to the poorest, achievement of the UN Millennium Development Goals will remain an unrealisable dream without the creation of sustainable economic opportunities for underprivileged people in rural areas. How do you create sustainable economic opportunities for poor people in rural areas? A good start would be to provide voice and data communication affordable for all.

Reliable, affordable access to telephony services is a requirement for virtually any realistic development scenario that aims to empower people to improve their lives by building local businesses based on creative thinking and hard work.

However, the business models applied by mobile network operators in developed markets to build their subscriber bases and revenues are, with few exceptions, inappropriate for the economies addressed by the Millennium Goals. Attempting to build mobile networks in rural areas using existing business models would result in parallel infrastructures and operations, which would negatively affect both environmental and commercial sustainability.

There are good reasons why telecom operators do not expand into rural areas.

- Many operators earn good money where they are today: in the cities, high density areas and along the highways.
- Using the current business models, they don't believe they can build communication for all in rural areas and maintain their profitability.
- Today, the operator has to carry all the business risk for both CAPEX and OPEX.

New business models are needed where all stakeholders can see profitable opportunities, including the consumers. So, without profitability for all, no sustainable solutions will be implemented. And for that to happen, financial risk needs to be shared among the stakeholders.

To summarise: the connection between profitability and sustainability must be understood by both the public and private side for a partnership between the two to work.

GIVE THE POOR A CHANCE

Being cut off from most kinds of communication keeps poor people in poverty. It is also very convenient for those who exploit the poor for their own benefit. One example of this are the middlemen who buy goods, crops, fish etc. from the poor. In most cases, they pay the farmers or fishermen too little for them to make a living.

How can communication help? Mobile phones can easily provide daily information on market prices, on crops or fish using SMS based applications. Making poorer traders more aware of how much they can get paid for their goods will reduce the power of middlemen and allow the poor to get better prices for their goods and thereby, be able to improve their quality of life. This was just one an example of many how mobile communication and locally tailored applications can help the poor to improve their quality of life.

According to studies (Ericsson Consumer Labs 2006–2007) poor people put great value on the access that a mobile phone brings. Having a number and becoming accessible opens up informal business opportunities for a large number of people. However, not much will happen unless we bring the cost in line with the local income levels. For a subscriber living on US$2 per day, a US$6 per month ARPU (Average Revenue Per User) represents 10% of their income.

Serving these markets cannot be about progressively lowering prices. It's about creating a new price-performance envelope:

Quantum Jumps in Price Performance are Required to Cater to the Poor Rural Markets

Providing mobile communication to the underserved areas in developing countries will bring much more to the table:

> 'A country with an average of 10 more mobile phones for every 100 people would have enjoyed a per capita GDP growth higher by 0.59%.'

This is regarded as a quite conservative figure; other studies show much higher GDP gains. An example of this follows, based on a study made by McKinsey.

By promoting the use of mobile phones, regulatory and industry players can amplify GDP gain depending on and coupled to mobile penetration. Comparing India, China and the Philippines, we can see clear differences in GDP gains. In India where the penetration was the lowest the GDP gain was 2%; in China with higher penetration the GDP gain was 5%; and in Philippines where the penetration was the highest among the three countries the GDP gain was 7.5%.

In India it was forecasted that a 10% increase in mobile penetration would generate an increase in economic value of US$2.3bn and further operator revenues of US$6.2bn.

Experience shows that operators and regulators can best spread the use of mobile phones by lowering the cost of owning and operating them. Simply cutting subscription prices could cut the operators profitability and hinder their ability to increase mobile phone ownership. There are many ways to increase profitability and, at the same time, lower the tariffs. In many emerging markets, subsidies on handsets are very high and have a major negative impact on the EBITDA. At the same time, the supply of handsets does not seem to be an issue. In fact, in many areas, low-income households sometime owns several handsets as a result of high subsidies. Operators could consider stopping the sale handsets or remove subsidies, thereby removing a huge negative impact on the EBITDA. This has to be done in conjunction with lowering tariffs dramatically. The result will be MOU (Minutes of Use) increases of several hundred percent, driving traffic growth and at the same time improving EBITDA maybe from below 20% up to 40–50%.

Another method is *the calling party pays* which is present in different variations in all of the three countries studied. In the Philippines, for example, only the calling party pays a connection fee, and it is free of charge for the receiving party. The Philippines improved the plan by offering free incoming calls for three months to mobile subscribers for each month worth of prepaid cards they used. This move, along with a favourable interconnection regime, has helped reduce the cost of owning and operating a mobile phone to an average of US$2.41 a month, the lowest level of the three countries studied.

Conditions within these markets vary considerably. Regulators should therefore tailor their actions, focusing on policies that optimise the operator's scale and the use of the network. Operators and regulators could, for example, promote network sharing to improve coverage and capital efficiency, especially in rural areas. Regulators can also set interconnection charges, taxes and activation charges at levels to help lower the minimum cost that customers pay.

Given the positive ripple effect that mobile phone penetration has on a country's economy, regulators and operators would do well to explore these and other novel ways to get more mobile phones to more people, especially in poor rural areas.

I think it is important to understand that commercial thinking and ambition to help the poor can go hand in hand providing that we work together and use the right business models.

Recent studies show that the poorer people are the bigger percent of their available income they are prepared to use for mobile communication. Poor people make tougher choices every day than people in the developed countries needs to do. One example of this is when faced with whether to buy food today or to top up the prepaid card, many of the rural poor choose to top up the prepaid card.

This is logical in their situation. Instead of paying for a 3–5 hour buss trip to buy products from the small retail shop in the home village and perhaps finding that there are no products available that day and going back empty handed, they can make a phone call and save the money for that trip. That is, if they had access to communication.

Another example is the farmer that used to buy pesticides for US$10 who can now make a couple of phone calls and buy the pesticides at half the price from another source.

One finding made from a study in Kenya by Ericsson: the proportion of income spent on mobile phone calls is higher among those with lower levels of income. Low income households in the survey spend 10%–40% of their income on mobile calls, with the majority spending around 10%–20%. People in the higher income segments of the survey spend around 2%–10% of their income on mobile phone calls.

Opportunities for Users

The value of information is often closely related to its immediacy and relevance, and in these respects the mobile phone is, in many circumstances, unrivalled. Access to information such as local weather and market prices for crops would, for example, greatly assist many poor farmers around the world who today live without such information.

Access to mobile communication saves time and money. In some areas of the world, it is often necessary to travel long distances in rough terrain only to discover the person, or the goods being sought are unavailable. One quick phone call or SMS could eliminate such wasted and costly trips. As bank facilities are not usually accessible for everyone in emerging markets, money transactions can instead be made using the mobile phone.

Mobile communication offers a wealth of opportunities for end users, from improved health care, security and better access to education, to farther reaching personal communication with loved ones and friends and access to information as well as entertainment. Most of these benefits are things that many people in the developed world take for granted, but for poor people in the developing world this represents a paradigm shift.

Opportunities for Society

Access to mobile and wireless technologies will help trigger sustainable economic growth in emerging markets. Through access to information (both voice and data) and improved communication, the burdens on society, in terms of poverty, corruption and disease, can be reduced. But it would be a fallacy to believe that these problems could be eradicated simply by providing mobile phones to people in these regions. Mobile phones should be considered an integral part of a larger package of measures needed to address issues such as health, education, law and order and emergency services. But if the solution is obvious on a theoretical level, why is this development not occurring today? Such development relies on investment – both in capital expenditure and operational expenditure, and also on the elimination of excessive government taxes and regulations, etc.

For operators and other investors, the return on investment has been unclear. A number of new stakeholders must be involved, such as governments, financing bodies and non governmental organisations (NGOs), if society is going to reap the benefits of mobile technology.

Opportunities for Operators

Mobile communication provides a means of communication that does not require the need to fully tackle the obstacles of literacy and language skills, which are required for the use of computers or other advanced personal devices.

Being able to communicate in one's own language is invaluable. Many small and medium sized businesses can be run using primary voice communication. Access to wireless telephony with voice as the basic service can therefore give rural and less developed regions the means to drive economic development from the bottom up. Computers require more resources on many levels and may, therefore, be an impractical solution, at least in the initial stages of providing communication for all.

Recent socio-economic studies by the UNDP and the Swedish Agency for International Development Cooperation (SIDA) in rural Tanzania have showed the following result:

The study showed that 97% of the rural population in the villages surveyed knew about mobile telephony and 50% had used a mobile phone. In contrast, 67% did not know what a computer was and only 3% had ever used a computer. Establishing Internet cafes in such areas would most likely be unprofitable. Evidently, there are institutions such as schools, hospitals and other governmental agencies that more or less immediately would benefit from PCs with Internet access in addition to regular voice communication.

CONCLUSION

Communications for all is not a myth. It is very possible to provide voice and data communication for all. An important word is *sustainability*, which goes hand in hand with profitability. Also, partnership between stakeholders is a very important enabler for implementing sustainable communications solutions. I believe that new solutions are within reach, but it requires that people change their mindset.

4

Customers' Needs for Telecoms Services and Applications

OVERVIEW

This chapter examines the behaviour and aspirations of users of mobile services and applications at the lower end of the consumer pyramid; consumers with income levels less than US$2 per day. We have also tried to analyse and evaluate whether these consumers use mobile phones in the same way as those at the higher end of the consumer pyramid[1]. Using evidence from Ericsson's Consumer Lab, the services that people in developing countries and at the low end of the pyramid demand are presented with simple examples. We have made an attempt to understand the communications needs of the consumers sitting at the lower end of consumer pyramid and how service providers have approached this segment in the past. We have also suggested ways in which service provides should change their go-to-market model towards these so-called *low-end* consumers. We also look briefly at the type of applications being used and deployed in the high-end segment and their relevance for consumers in developing or under developed markets and living in remote rural areas.

The provision of service is not without challenges and we take a look at some of the challenges associated with penetrating a market with advanced mobile data technology and services which includes the

[1]Consumer Pyramid is the way one can segment the market place. Consumers with high income sit at the top of the pyramid while the consumers with lowest income sit at the bottom of the pyramid.

Business Models for Sustainable Telecoms Growth in Developing Economies
S. Kaul, F. Ali, S. Janakiram and B. Wattenström
© 2008 S. Kaul, F. Ali, S. Janakiram and B. Wattenström

business case for deploying advanced mobile technologies and service enablers, access to smart handsets and personal computers, maintaining charged batteries, the fear of technology among consumers, creating and maintaining local content relevant for the these users.

UNDERSTANDING THE NEEDS OF THE LOW-END CONSUMER SEGMENT

Historically, the telecoms industry has not cared about the needs of consumers at the bottom of the pyramid. In developing nations, consumers had to wait in long queues even to get access to basic telephony. From my childhood in India, I remember that when my parents wanted to have a telephone line at home, they had to wait for about six years before they could get one. Things have changed since then and mobile technology that has made this possible. However the approach to the market has consistently been inside-out; as a mobile operator you decided what you wanted to offer in terms of products and services, you chose the technology that could enable them and as the technologies evolved you decided to offer new services like MMS, email, voice messaging, video messaging, video telephony, internet access etc. With further advancement in technology, mobile operators started offering high speed broadband services.

For developing nations this is truly what I call 'moment of truth', the dream of bridging the digital divide seems to be close to fulfilment. The consumers deprived of access to information now seem to be getting closer to achieving it.

The rural market has distinct characteristics. These characteristics must be understood in order to define clearly the market segments that exist and the needs of these segments.

It is a fundamental mistake to assume that the low income rural market is not viable. If we take a look at aggregate purchasing power in developing countries most of the people exist on incomes of less than US$2 a day. However, the developing countries *put together* offer tremendous growth potential. It is safe to say that the next 3.5 billion mobile users reside in China, India, Africa, South and Latin America, Russia, Pakistan, Indonesia, Thailand etc. Collectively, they are home to future growth in mobile telephony, and hence, can't be ignored. Yes, it also needs to be understood that the consumer in the developing nations doesn't have the same purchasing power in dollar terms as the consumer in the developed nations. The assumption that is that consumers with less than US$2 income per day cannot afford advanced communication services and applications. However by the mere virtue of their numbers, we should make these advanced services and applications available to these users and drive the business case by economies of scale. For example, making high bandwidth Internet access service available to these users would be the only option for them to get faster access to Internet. During a recent survey conducted by the BBC to analyse WAP traffic, it was noticed that the majority of the traffic came from Africa and few other developing countries.[2] The logic is simple: the consumer in the developed nations can get access to news broadcasted by BBC through number of different means, but for consumers in rural areas in the developing nations, mobile WAP access is possibly the only means to access BBC broadcasted news.

Through our research in a number of developing countries, it was revealed that even though the tariffs were lower compared to developed countries, *the average spend*[3] by a consumer in the developing countries is a much higher proportion of their net income. In fact, it was also found that lower you go in the consumer pyramid, the higher people pay per minute of usage. The lower the denomination of the prepaid voucher, the higher the pay per minute / per megabyte of use.

Most of the developing nations still face lack of Internet penetration and the fundamental reasons is the unavailability of telephones lines and computers. Speedy mobile penetration has solved both the issues, mobile connections are faster and cheaper to deploy and the mobile phone is the first computer experienced by many rural consumers.

To make the technology and the services and applications it can enable relevant for consumers in rural areas, mobile operators and service providers need to understand these consumers. How do you do this? The consumers sitting in the bottom of pyramid apart from being poor are not connecting, hence understanding who they are and what they want is difficult.

The mobile phones have provided a lifeline for most of the users. The consumers in the rural areas in the developing economies are using this tool to improve their livelihoods. Our *research*[4] in sub-Saharan Africa revealed some interesting facts about the key benefits of using mobile communications by low income groups.

DO PEOPLE IN DEVELOPING MARKETS USE THE PHONE IN THE SAME WAY AS IN DEVELOPED?

To get the answer to this question, our research team have been analysing different developing markets around the globe. The results more or less

---

[2] http://news.bbc.co.uk/1/hi/world/africa/4795255.stm
[3] Average spend means communications spend as a percentage of total spend.
[4] Ericsson conducted research in a number of developing nations to understand what the consumers of mobile services in these countries think and want.

match the finding of Ericsson's recent study in sub-Saharan Africa. The key themes of why consumers in the low end market segment of the developing countries, the so-called next 3.5 billion users use mobile communications services are:

Theme 1: Generating & Improving the Income Levels

The primary benefit of the mobile phone is its positive impact on earnings. With mobile phones people are able to run their businesses more efficiently than they would if they had to rely on personal meetings. Instead of travelling to suppliers – which uses up both time and money, as well as requiring the business to close temporarily – the mobile phone can be used. Furthermore, it is possible to make direct contact with clients, which in turn generates more business.

The benefits are dramatic and clearly evident, creating a huge interest among non-owners in obtaining a mobile phone: people with mobile phones actually earn money! The study has uncovered numerous examples showing that each call can be an investment that pays off in the end.

Smaller than Microbusiness: Nano Business

"Get money" and "Save money" are used as synonyms for "Earn money". People often talk in terms of saving money, but in reality they refer to earning money.

Several of the self-employed businessmen and women in the rural villages and the low income areas of Nairobi run very small scale enterprises with daily turnovers of approximately Sh1000–5000 (US$15–70). Competition is fierce as there are numerous businesses all selling the same goods. It is therefore difficult to make ends meet. Every deal is crucial and the mobile phone makes it possible to create and maintain customer relations.

The Potential Positive Effects on Earnings Make Those in Low Income Segments Prospective Mobile Phone Users

Self-employed people and casual workers have most to gain from owning a mobile phone. A common saying among the people who took part in the study is 'you have to use money to get money' (i.e. you have to spend money to earn money) meaning that buying a mobile phone and paying for top-up cards are considerable expenses, but they are expenses which can create greater earning potential.

Fewer Expensive, Time Consuming Journeys to the Supplier

Fredrick has a small general store where he sells flour, sugar, paraffin, cooking fat and bread. Before he got a mobile phone he had to close the shop for at least one hour in order to walk to his supplier and pick up more supplies. Today, he spends Sh10 on a mobile phone call, has the items delivered and keeps the shop open, meaning that he won't lose any business. Furthermore, he calculates he will make a profit of Sh90–100 on selling maize meal after using Sh10 on a mobile phone call to order three bags of it.

Making Plans for the Day is Easier with a Mobile

Rogers works for the council during the day, and owns a kiosk where his cousin sells shoes. Rogers calls his cousin very early in the morning to discuss which shoes they need in the shop. He then calls the supplier and tells him that the cousin will come and collect the required shoes before the shop opens. The supplier can also call Rogers and inform him when the shoes that they know he is interested in are in stock. The cousin gets the shoes on credit and Rogers pays a visit to the supplier after work to pay for them. Without his mobile it would cost Rogers Sh100: first to take the matatu (bus) during the morning rush hour to talk to his cousin; then to visit the supplier himself and deliver the shoes to the kiosk.

Developing Nano Business

Before, when I needed to make an order from the suppliers or the wholesalers, I was forced to close the shop and walk up to their place or work to get the goods. But now I just make a phone call. It is saving time because I am not closing the kiosk to go and make an order. — To give an order of three bags of maize-meal will cost me just Sh10 and from the three bags I can make a profit of Sh90–100. So by the end of the day I will have saved something.

Fredrick, 20 Nairobi

'I can actually earn money from using my mobile phone. That's why most of my calls are business related. I know that even if I am going to spend Sh100 that at the end of the day it is going to give me Sh1000 back. Without the stock I would not get the Sh1000.'

Rogers, 37, Nairobi

Spending Time on Work That Brings in Money

Milka has been a single mother since the death of her husband. She earns her living from farming and has a shamba in Gikoe where she grows tea and coffee to sell. She also rents a shamba 4 km further down into the valley where she employs casual workers to harvest the crops. As she does not have a mobile phone and the Semi ya Jamii (communal phone) is situated 4 km away, she has to walk to her rented shamba twice a week to find out how things are going, instead of just being able to make a phone call and spend the journey time on her home farm instead. Milka wants the mobile network to include the valley in which she lives – then she will save up to buy her own mobile phone and use the time saved through having this to make money and cultivate her own shamba, instead of wandering backwards and forwards, up and down the valley.

More Business for the Barber

Simon runs a barber's shop in Nairobi together with his cousin. Business is up and down, but has become a little more predictable since his mother bought a mobile phone two years ago. Since then, customers have been able to call him on his mother's mobile and make appointments for haircuts or shaves; Simon can also call his customers and change appointments when necessary. All of these factors mean that Simon can plan his days in a more efficient way. Before having the use of a mobile, he was never sure of whether he would have any customers during a day. If he then closed the shop to take care of some other business, customers could turn up only to find the salon closed – resulting in decreased revenues.

No Need to Close the Shop to Get Supplies

Nancy and her husband – who works in Nairobi – set up the store in Kianganini four years ago. During the first year she did not have a mobile phone and had to close the store for a day and pay Sh400 take the matatu bus to Muranga whenever she needed supplies. Nancy's husband had a mobile phone for his work in Nairobi, and the couple began to forsee the benefits of a mobile phone for their business in Kanyanaini, deciding therefore to postpone the purchase of a goat and buy a mobile for Nancy instead. Nancy now calls her wholesalers' mobiles when she needs more goods to be delivered. She can also call and place an urgent order if she runs out of stock. Furthermore, Nancy sells top up cards in the store for Sh4000 per week – a nice addition to total turnover. Regular customers who, for one reason or another, cannot make it to the store in person, can call Nancy and ask her to Sambaza (transfer) some airtime to them, which she promptly does and receives payment later. Nancy also uses her mobile phone to call the bank and find out her current account balance, in order to decide how much money's worth of goods she can buy. She is also able to discuss investments with her husband in Nairobi (which she would not previously have done from a public phone booth, due to the lack of privacy). The mobile thus helps her to budget and make good business decisions.

Taking Better Care of the Cow Provides More Litres of Milk to Sell

Julius is teacher and farmer in Kanyanaini. A mobile phone allows him to be able to make calls from home, which in turn enables him to look after his cow properly instead of having to go all the way to Nairobi just to say that he is unable to attend a meeting – which was precisely what he had to do before he got a mobile. Taking better care of his cow provides more milk to sell and brings in more income for the family.

Take Orders Over the Phone

Jemi has a beauty shop where she sells necklaces, bracelets and nail polish. She also does manicures and her mobile phone is extremely important to her as customers often call to make bookings or to ask her if she has the latest nail polish colours, or a necklace they saw in Nairobi but did not have the money to buy it at that time. Jemi calls her suppliers, places orders and calls her customers when she has received the desired products.

Close Relations with the Customer Means More Business

George shares a small stall with a friend and both of them sell clothes. If customers do not find what they looking for they will place an order and George goes to the market to get the wanted items. He uses his mobile phone to contact his clients, who either will come and pick up the objects or ask George to deliver. Either way, George can be sure he will have a deal and the cost for making the phone call will be covered by his revenues. Before he got a mobile he had to use the phone booth or use his partner's mobile. The phone booth was expensive and after Safaricom[5] started charging for diverts from his line to his friend's phone he stopped diverting his calls and started receiving calls on his friend's number. But he didn't feel at ease making use of his partners' mobile and saved Sh300 per week for a phone of his own (in the end he actually got a phone from another friend).

Other people in the study, like Fredrick, agree that the mobile phone brings more business due to better connections with the customer.

Help Finding Supplies

Nancy is a dressmaker and is always on the look out for new fabrics and materials. It is crucial for her business to find the materials and colours that her customers ask for. She often visits the market in Kangema, 8 km (and Sh80) away from where she lives, in order to find fabrics. The supplies at the market vary and Nancy can never be sure that she will find the fabric she is looking for. Her brother, who lives in Kangema, helps her out by checking fabrics and materials on each market day. Prior to embarking on a trip to the market, Nancy calls her brother: if he says that there are no interesting fabrics then Nancy will not make the journey. The call saves her both time and money. Nancy also helps her brother – he is a

[5] Safaricom is one of the leading mobile operators in Kenya.

carpenter and is constantly in need of timber, therefore Nancy keeps an eye out on his behalf. When she gets hold of some timber or finds out where her brother can make a good deal, she calls him.

Check Prices and the Amount of Tea Sold at Auction

John has two sources of income: he is employed as a vet by the town council, and he cultivates his relatively large shamba[6], from which he harvests tea to be sold at the tea auction in Nairobi. The mobile phone has helped John to check up on his tea at the auction in Nairobi. He now makes a call to see if his tea has been sold, find out how much he was paid for it and when payments will be made to his account. As tea is sometimes left aside and not auctioned, John finds it important to be able to check on his harvest. When he did not have a mobile, he went to the auction by matatu, so he now saves both time and the Sh800 ($11) used on bus fares.

Save Money on Cheaper Goods from a Larger Town

Puriti's family has recently experienced a drop in income since her husband was laid off from his job at Barclays Bank in Nairobi, two years ago. Puriti's husband is still in Nairobi doing casual work while Puriti and the children live in Kanyananini, where Puriti works as a teacher and takes care of their animals and small shamba. They earn some extra money through selling tea and eggs. As it is a problem making ends meet (especially when it is time to pay the school fees), Purity tries to save money in order to increase earnings wherever possible. One way of doing this is through using a mobile phone to order fertilizers for the shamba. Fertilizers in the larger town of Kangema are Sh50 cheaper per bag compared to the prices in Kanyanainin where Puriti lives. She therefore calls a distributor in Kangema when she wishes to place an order, and the goods are delivered to her door. By doing this, Puriti saves Sh50 per bag which – when buying five bags – leads to a saving of Sh200, as the phone call usually costs Sh50.

With the Mobile There is a Chance of Finding a Job for a Day or Two

James and his wife have a small shamba, where they live with their two daughters, a cow, a calf and seven chickens. When James bought a mobile

[6]Shamba is a small plot of land.

phone for the money earned from last year's coffee crop, plans to improve the family's corrugated tin shed were postponed. The family earns Sh1200 monthly and casual opportunities for James to work as a matatu driver are extremely important – which is why he bought a mobile phone. James always needs to have Sh30 of air-time, to be used to call back when an employer has sent a 'call-me' message. When money is scarce it is hard to choose between buying food or a top-up card. However, despite the strained situation the possible chance of getting a casual job makes life with a mobile phone more tolerable than it used to be without one. Today he might get a job after being idle 3 days while earlier, when he did not have the mobile phone, he would be out of work for 3–4 months at a time.

Find Casual Work and Employment

Peter used to work as an agricultural adviser but was retrained two years ago. He now works as a voluntary teacher within the Catholic community and takes on casual work during school holidays, which is very important in order to be able to make ends meet. For Peter, the mobile phone is invaluable as it helps him in the process of searching for jobs. With his mobile he contacts people within his network (friends, former colleagues, former employers) and asks them to look out for casual work on his behalf (riding errands on his motorbike, harvesting, logging trees etc). With the mobile, Peter is always accessible and can start work at very short notice. When he receives information about a job opportunity, the mobile phone makes it possible for him to gather more information and arrange interviews.

Find Casual Workers and Keep up to Date on Volumes at Shamba

As a casual labour employer, John can see the many benefits of a more widespread use of mobile phones. As John is in full-time employment (as a vet), it is primarily his wife who takes care of the shamba and employs casual labour to help with the harvest. As the majority of the labourers they employ don't have mobile phones, John's wife travels to former workers' homes and leaves messages when help is needed. Sometimes workers also come to the house themselves and ask if they can help. On some days these workers are lucky but on other days there is no work and they have to leave to search for another job, meaning that they have wasted time and possibly missed other job opportunities that day. John finds this arrangement inefficient. If more casual workers had access to mobile phones all involved could make arrangements over the phone, saving time for everyone concerned. John could also communicate with the labourers on the shamba and check how much tea there is to harvest and how much of it has actually been harvested; thereby allowing him to hire more workers or lay off workers depending on the amount of leaves that are ready to be harvested on that particular day. He currently has to take a day off from work or work shorter hours in order to check up on the status of the shamba. If at least one of the casual workers had a mobile phone it would allow John to stay at the office or carry out other tasks at the farm.

The Mobile Creates Work and Business Opportunities

Simon is employed as a mobile phone retailer in Kangema. He started the job two years ago, and was tremendously excited about it as he had been unemployed ever since finishing secondary school. Simon is enthusiastic about all the benefits to society and consumers that network coverage and mobile phones has brought. He even got his own job due to the expansion of network coverage and mobile phone services. There are so many other businesses that have also developed, for example the Semu ya Jamii business, sales of scratch cards, mobile repair services and charging services. In his daily work Simon meets a lot of people who want to buy mobile phones, as they are convinced that a mobile can help them generate more business and income.

Theme 2: Creating and Strengthening Social Networks

Helping each other is part of the Kenyan culture and networks are of crucial importance for survival. The absence of a social security system makes it necessary for low income households to build a network based on relationships with family and friends. Staying in regular contact is important for maintaining relationships. Frequent contact also creates a feeling of security in the knowledge that the network exists. Everybody knows that helping somebody else today means you can get help yourself tomorrow.

    Most phone calls take place because of a basic need for support. The social networks serve as a hub for finding work, education and business deals. The main reasons for the majority of mobile phone calls within low-income segments are to obtain information or money, to check on a friend's or relative's needs, or to give information about a job or business opportunity. By using the mobile phone, contacts can be more frequent and information on job opportunities or education can be forwarded a lot faster. Consequently, the social network grows stronger.

The Mobile has a Central Role in Helping One Another

It is often money matters that network members help each other with, and at the same time that the mobile culture starts to reach more people, the phones are increasingly being used to transfer airtime (sambaza). When somebody has a few shillings to spend it is often used to sambaza a friend or close relative a a small token for closeness. As a result of the useful sambaza service there is a great openness towards transferring money directly between mobiles.

Lack of money and an insecure future create a huge need for support.

The mobile phone is also used to offer support when people are down. Young people in their 20s, who come from families without the resources to pay for higher education, feel that they will never be able to create a better future for themselves and their families. Education is the key to employment and tolerable living conditions, and these young people simply do not have the means to be part of this better future. They feel stuck where they are and have a strong need for encouragement and support.

Tradition also adds to the frustration: the first-born in every family is expected to help out on the farm or in the small business while younger siblings are able to attend higher education.

The Community Mobile Phone Makes it Easier to Ask for Help

Joseck lives in Gikoe with his mother and takes care of the shamba while his mother does casual work at a neighbour's farm. Joseck's father died a couple of years ago and as the family's income is very low and their earnings from coffee are only paid out once per year it is always a struggle to make ends meet. Assistance from relatives is always necessary. As Joseck lives in a valley without network coverage, he sends letters to relatives when in need of monetary help. A letter can take up to two weeks to reach the recipient, so when things are really urgent, Joseck walks to the town of Gikoe 4 km away where he uses the Semi ya Jamii.

Sharing What We Have

Fredrick often lends his mobile phone to friends who don't have any credit and need to make a call. He knows that next time he will probably be the one with no credits needing to make a call. George lets his closest friends put their SIM cards in his mobile phone to make a phone call or text. He used to do the same before he got his own phone and wants to help his friends as he was helped himself.

With a Little Help from Your Friends

Emmah gets airtime from her boyfriend and her brothers who all have a better income, compared to hers. They sambaza her every week, enabling her to spend Sh400 on calls to them. She beeps her boyfriend and brothers, but while in Nairobi, she would not beep her mother in Kanyanini, because her mother would then have to walk to the shopping centre to get airtime to call her.

A Family Phone is of Great Help in Business

Lizzy is a business woman selling second-hand clothes. She doesn't yet have her own mobile phone, although borrows her mother's whenever possible, as it helps her immensely in her business. On certain days, Lizzy's mother takes her mobile phone with her to work, leaving Lizzy without access to it. If Lizzy has to make important calls during these particular days, her cousin lets her borrow his mobile phone. Lizzy then purchases a Bamba 50 (Sh50 top-up card) and tops up her cousin's mobile before making her calls.

Encouragement Via the Mobile Phone

Both of Jemimah's parents died a year ago and her sister is ill in hospital. Jemimah lives together with her brother, second sister and her sister's baby, on their inherited shamba. She also runs a small business in the town centre. The last few years have been tough both economically and emotionally, and it is sometimes difficult to find the will to carry on. When feeling down and short of money for school fees, hospital fees, food or communication, Jemimah knows she can get help and encouragement from friends and relatives. They always help each other, either by giving money or sending a sabaza to at least make it possible to communicate and receive encouragement when times are difficult.

The Mobile Offers Hope

Elisabeth's visit to her older sister in the village was involuntarily extended from a few days to several months. Elisabeth has to look after her niece and nephew while her hopes for a future seem to be going down the drain. Elisabeth wants to go to college in Nairobi but her father has told her that he can not afford it. She can see no possible solution to this problem and feels that her chance to make a better life is evaporating before her. She walks to the Simu ya Jamii three times daily to talk to her favourite sister, and borrows her older sisters' phone at night, when she is back from work, to talk to friends. The communication offers her some relief when her favourite sister and her friends give her encouragement.

Theme 3: A Modern Infrastructure That Works

The development of the infrastructure in many developing nations has been neglected for many years, resulting in a situation where many villages and households do not have roads, electricity, water or access to landline phones. A typical situation today is that electricity and landlines are present in the commercial centre of a particular area, whereas the villages surrounding this commercial centre have no access.

Phone booths have, up until recently, been available in most town centres, but have been unreliable as they are often vandalised or out of service due to poor maintenance. Many interviewees describe long walks to the phone booth, only to discover that it is out of service.

Letters are an alternative means of communication, although they travel very slowly in rural areas, taking up to two weeks to reach the recipient, compared with a mobile phone call which is instant.

The Mobile is Perceived as the Major Infrastructure Development of Recent Times

Mobile network coverage has offered consumers in rural areas the most rapid infrastructure development they have ever experienced. In a situation where roads are poor or nonexistent, and when there is no electricity or landline phones, the mobile phone connects people and delivers information which previously took a long time to come through. The possibilities offered by the mobile phone can considerably enhance living conditions for people.

All Contacts Benefit from the Mobile Phone

The positive effects from using the mobile phone are considerable in areas such as:

- Communication – Less travel and no need for messengers
- Heath care – Access to medical centres, doctors, information, medicine etc.
- Education – Up to date information and communications with schools and teachers
- Finance – Access to bank account balance, information and services (electricity bill)
- Security – A way to reach police or other forms of help quickly
- Corruption – Convey inappropriate actions made by officials

More Reliable than Messengers and Postal Services

As sending letters is so slow, Joseck often turns to a messenger when wanting to get in touch with a relative living elsewhere. He gives a message to someone travelling to the particular town or village were his relative lives and pays a small fee for the service. This way, the letter reaches the receiver quicker than it would through ordinary postal services. This is, on the other hand, an unreliable method as it is up to the discretion of the messenger as to whether the letter is delivered or not. Privacy can also be an issue here, as the messenger may read the content. Joseck would really like to have his own mobile phone and intends to save up for one as soon as there is network coverage where he lives.

A Lot More Efficient Than Letters

Before Nancy and her siblings owned mobile phones they used to send letters to each other via the church. It could take up to two weeks for the letters to reach their recipients. Nowadays, they prefer to call each other as this is so much quicker.

'Please Call Me' is Used as a Coded Message

Mobile service providers offer free 'Please call me' messages (beeps or flashes) that between friends are used as coded messages to say 'I have arrived safely at home', or 'I have arrived at the place we decided'. They are also used for the intended meaning, however the sender normally is very cautious not to send such a message to somebody who would be worried and go through a lot of trouble to find air-time to call them back.

Getting a Diagnosis Over the Phone

Jemimah has regular contact with the health centre. She lives with her sister's 5-year-old son, and as children often get sick, there are plenty of situations when a doctor is needed. The mobile phone is of great value for the household in making this type of communication. Jemimah can call the health centre and get a diagnosis which, if the problem is not serious, saves her the time and the expense of a trip to the clinic on the matatu bus.

Making an Appointment

Nancy also finds the mobile phone useful in her communications with the health centre. When she or her children are sick she always calls the health centre in advance to check if the doctor is there and to make an appointment. Before she owned a mobile phone, the family had to travel 8 km to the public health centre (cost Sh200) or pay to see the doctor at a much more expensive private clinic in Kanyanaini.

Checking if Medicines are in Stock

John takes care of his ageing dad who requires daily medicine. John uses his mobile phone to check if the public health centre has his dad's drugs in stock, as it is very common that they are not. John still remembers when he had to travel to the health centre only to find that the medicine he needed was out of stock: a trip that was both time consuming and expensive.

Easier to Get Educational Information

Puriti uses her mobile phone when evaluating collages for her daughter. In collecting information about enrolment criteria and fees for different schools the mobile phone is a great help. Instead of writing letters (which may not even reach the school in question), or travelling to each and every faculty, the mobile facilitates efficient communications which often cost less than a bus fare.

Information on Time

Jemimah's old school holds occasional seminars and invites former and current students to participate. The invitations used to be sent out in the form of a letter, but are now issued via a phone call. Jemimah appreciates this a lot as the letters often arrived when the time for registration had passed, making it impossible for her to participate.

Information About Parents' Meetings

Before the mobile phone arrived, information about parent's meetings would be sent via letters. However, more often than not, the meeting would be long past by the time the letter finally arrived. Emmah, who studied in Nairobi while her parents lived in Kanyanaini, points to better relations through improved communication with parents living in rural areas as one of the main benefits of the mobile phone compared to sending letters.

An Efficient Tool for the Collegium[7]

Julius works as a teacher in Kanyanaini, where several of his colleagues have bought mobile phones during the last year or two. Julius finds that using mobile phones has helped teachers to work more effectively compared with previously, when they had to walk to attend various

[7] Colleagium is any association with a legal personality.

home-visits (or indeed, walk to the school) to pass on messages, send letters that would normally take weeks, or to avoid using messengers who could potentially distort a message. Now, the headmaster and teachers just have to call each in order to distribute information. The positive benefits of the mobile phone have made Julius encourage all of the other teachers at the school to buy a phone. Consequently, most of the teachers now have mobiles.

Easier Contact with Teachers at School

Nancy uses her mobile phone to communicate with her children's teachers about different issues. She calls them to ask how the children are doing at school or whether new books are needed in the near future. This information makes it easier for Nancy to plan for upcoming expenses. Before she had a mobile phone, Nancy had to close her shop for 3 hours while she walked to the school and spoke to the teacher, hence losing income opportunities.

Calling in the Father of the Family

When there is a need to talk seriously with the young boys who are running wild or are on the verge of dropping out of school, a mother needs a father to be present. Julius is teacher at the school and is happy to lend out his own mobile phone to a mother so that she can call the father of the family in Nairobi and ask him to come back and put pressure on the child. To Julius, not lending out his mobile would be saying no to the child, and because the safeguarding of future of the children at his school is his mission, he does not ask for any reimbursement for the calls made. Prior to being able to use the mobile phone, the only way to get hold of a father would be by sending a letter, which could take several weeks, allowing the youngster in the meantime to get into even deeper trouble.

Easier to Check the Bank Balance

John gets paid once a month for the tea he sells at the auction in Nairobi. As he needs to confirm that the correct amount has arrived in his bank account, he regularly checks his bank balance. To do this he calls his bank from his mobile phone, which is very easy. Previously, he had to spend Sh200 to travel by matatu bus to his bank branch, 8 km away.

Negotiating Terms

When Peter was dismissed from his job he had a bank loan which became difficult to repay while he was out of work. He occasionally has to renegotiate the terms of his loan and uses the mobile phone to call his bank manager. Previous to having a mobile, Peter was forced to take the matatu bus 8 km to reach his bank branch, which cost more money than the phone call.

A Way of Transferring Money

Julius is able to sambaza money to a teacher at his daughter's school, who then gives the money in cash to the daughter who does not have a bank account of her own. Prior to this option, Julius had to travel all the way to Nairobi in order to give his daughter money.

Call the Police When Something Happens

Rose lives in a village without network coverage but would very much like to have coverage and a mobile phone. The main reason for this is that it would help her to make the income from her own shamba and from casual work on other farms last longer. It would also be easier to communicate with relatives and friends. Rose points out that one would feel safer with a mobile phone as a lot of thefts and assaults take place in the evenings and at night, especially in the days following the once monthly milk payment when everybody has cash in their houses – something which attracts criminals. With a mobile phone, Rose would be able to call the police if something happened to her family or her neighbours.

Security Instead of Meat

Julius was once attacked by robbers in the period prior to his being transferred to his job at the school in Kanyanaini. At that time, he did not have a mobile phone because there was no network in the area where he lived. After moving to Kanyanaini – partly for reasons of security – Julius bought a mobile phone. Not long after buying the mobile, Julius' friend and neighbour called to say there were some people – possibly robbers – outside his house and he ought to call the police. Today, security is very important to Julius and he is very particular in ensuring that he always has credit on his phone in case of emergencies. Julius will often top up his mobile phone rather than have cash in his pocket or buy meat – he thinks it is better to buy less expensive goods and have airtime in case of an emergency.

Counteract Corruption

Daniel does not have a mobile phone, due to poor network coverage. He would very much like to have one though, as he believes it could boost his business. Furthermore, he views the mobile as a useful tool in counteracting corruption. If Daniel had a mobile phone he would be able to call the radio station if he knew that something corrupt was taking place within the council authority where he lives. Daniel listens to the radio a lot and once heard a report about a man who called in to the radio station from his mobile and revealed that the local police chief was involved in corrupt behaviour. Following the report, an investigation was launched and the police chief was ordered to leave his position.

Theme 4: Enabling Direct Contact and Privacy

Direct Contact

The mobile phone enables people to get in direct contact, without having to rely on communal phones or messengers to forward their messages. The difference from how it used to be is dramatic, according to everyone who participates in the study.

Calls are Often Confidential

As phone calls within low income segments are mainly used for important issues concerning monetary assistance, investments and private family matters, privacy is a great concern. One advantage of having access to a mobile hand set (owned, family phone or friend's phone) is that it offers privacy and enables direct contact – benefits unavailable when using a public phone booth or Semu ya Yamii.

Privacy is a Luxury

Family members in low income households live together in very small houses, making it difficult for everyone to have their own space and to do things without everybody else knowing about it. The compact living situation creates a build up in the need for privacy among people – a need that the mobile phone can satisfy. Privacy is of particular importance to young people on all continents and in all countries, and the mobile phone has made it easier for everyone to have a life of their own.

Direct Contact Saves Time and Money

Almost everyone in Felix's network has access to a mobile phone, making it easy to get in direct contact with each other. The mobile phone has saved them all a lot of time, inconvenience and money. Getting in contact with relatives and friends was much more difficult before the mobile arrived. Felix remembers the time when neither he nor his grandmother had a mobile. He would use the phone booth to call her neighbour, who he then kindly asked to go and fetch his grandmother. He would hang up and call back after 20 minutes, hoping that she was at home when the neighbour delivered the message – which was not always the case. Another thing that happened to Felix was that he once travelled to a friend's house, only to find that the friend was not there: a trip which turned out to be a complete waste of time and money.

Contacting Extended Family on Special Occasions

Jovinne finds his mobile phone very useful on special occasions when distant family and relatives must be contacted. When people live in different towns or villages, keeping in touch and giving each other updates on family matters is important – especially if someone is ill. Without mobile phone access this communication takes time; which is a big problem when someone has passed away, as attending funerals is extremely important to Kenyans. Jovinne explains that when someone has died, the closest family member must get in contact with all relatives and invite them to the funeral. With slow postal services this can be very difficult, and the advantages of mobile phone access in this situation are very evident.

Long-Winded Contact Routes Without a Mobile

Siphora studies in Nairobi. Her mother has no mobile phone and if Siphora wants to talk to her she has to call the Semu ya Jamii woman in town, who then sends a messenger to Siphora's mother telling her to walk the 4 km into the town and wait for Siphora to call. Personal messages such as these, however, are often distorted or incomplete: Siphora's mother has even spent a whole day waiting for a call because she only got to know

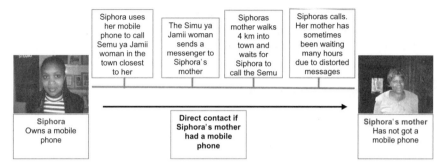

Figure 4.1   Long winded contact route without mobile phone

the day on which Siphora would call, not the time. These problems are illustrated in figure 4.1 below.

Phone Booth Queues Limit Privacy

John often calls his son in Nairobi to discuss investments on the farm. The mobile phone has made these discussions easier as it offers privacy. John remembers the time when he did not have a mobile phone and had to use the phone booth. People were always queuing, waiting for their turn, which restricted John when talking about money and investments, as everyone could hear what was being said. For John, money matters are – by nature – private, and as crime rates are high there is always a risk of break-ins if people have reason to believe that somebody has money.

Semu Ya Jamii Operator Comes to Know a Lot

Jane uses the Semu ya Jamii in the commercial town centre of Gikoe when she has to make a call. The Semu ya Jamii operator can listen to everything that Jane says, which doesn't feel particularly good, especially when she is asking for money or discussing other private family matters. She would like to have her own mobile phone, as that would offer her more privacy.

A Family Phone is Not Enough

Simon would like to have his own mobile phone. When his mother bought a mobile, his life became easier, although he would like to have the privacy offered by having his own. His mother now answers almost all of his calls, making it impossible for him to have any secrets. She is often close by when he talks to his girlfriend in the evening.

"It is nobody else's business how much we plan to invest in our farm. If people know that you have money you will get robbed."

John, 50, Kanyanaini

Borrowing a Mobile Phone Means Others are Eavesdropping

Before Emmah got her own mobile phone she had to borrow one from a teacher if she was in school, or from somebody else that was around if she was outside school. The reason for calling was often private matters that she did not want other people to hear, and it felt uncomfortable when the person lending out the mobile phone was standing around, waiting for her to finish. Besides, it could be difficult to get teachers at school to lend her the phone because they were very strict about who she was calling. With her own phone she is free to communicate to anyone she wants, whether it is friends or relatives.

Theme 5: Symbol of Class and Modernity

Network Coverage is a Sign of Development and Modernity in the Village

At village level, network coverage is perceived as a symbol of development. It shows that the village is part of a modern society where people have the same communication opportunities as other people who live in more urbanised areas. Thus, a base station with a visible mast is viewed as a symbol of positive development in the area, and as something which brings power to the whole region.

A Status Symbol: to Belong, to Not Be Among the Absolute Poorest

When network coverage and mobile phones first came to Kenya in 1999 it was only affluent business people who owned them. The mobile phone then penetrated the upper and middle class segments during the years which followed and became a symbol of prestige and status. Within low income segments, owning a mobile phone is still regarded as a sign of prosperity, which is a strong trigger for buying a device.

- 'I am not a nobody'

The Mobile is an Asset That Can Be Pawned

The mobile represents a monetary value that, in a situation of extreme financial difficulty, can be pawned and converted into cash. This option is, however, considered to be a very last resort to be used when all other alternatives have been exhausted, as letting go of the mobile automatically means less scope for generating income. This possibility is, nevertheless, perceived as a form of security.

The mobile phone can also be used as financial security with suppliers – even for just a few hours until goods have been sold and cash can subsequently be paid.

Boosters Everywhere Shows Development

When Emmah, who is visiting her parents in Kanyanain, is asked which areas she would most like to see developed in the region, she mentions electricity, mobile networks and hospitals, together with colleges for families who don't have relatives in the cities and cannot afford to pay boarding fees. Such developments would make the village of her birth more modern, and she herself would not have to go all the way to the shopping centre in order to charge up her phone battery. In addition to this, she would be able to talk from her bed, instead of having to go outside to find network coverage.

People Back Home Feel They are Living in the Modern World

Rogers explains the difficulties connected to making phone calls in the old days when the only phone booth was a kilometre away from his old mother's house: calling from Nairobi using a code number drained his money and his mother had to walk a long way to wait for his call, and as a result calls were scarce. With the mobile phone Rogers and his parents can call each others more frequently, and his parents are updated about everything happening in Nairobi. According to Roger this makes them feel more relaxed because they know all is fine with him and his family. The downside is that Roger's visits to his parents are reduced, but the number of calls made has definitely increased. Roger is certain that making use of the mobile phone makes his parents feel they are part of a modern world, despite their remote location.

Belonging to Another Class

Purity bought the family's first mobile phone four years ago. She realised that the mobile would make it easier for her to communicate with her husband who worked in Nairobi, although another strong argument for buying it was that it would send out the signal that she and her family belonged to another class; a more affluent class. As the family did not have enough savings to pay for the phone, Purity used money from a merry-go-round in order to be able to pay for it.

The Mobile Phone Shows I Am Not a Simple Man

Daniel is a businessman who lives in Nairobi. He bought his mobile phone 5 months ago as he wanted to use it in his business to generate more income. Apart from being useful to his business, Daniel admits that another reason for buying the mobile was that it brings prestige and shows that he doesn't belong to the lowest class.

A Mobile Phone Makes Others See You Differently

Emmah is certain that her mobile phone causes others to look upon her differently – i.e. as if she is somebody who has a lot of money. This is really not the case though, as Emmah was given the mobile phone by her boyfriend. Emmah also receives top ups from her boyfriend (who is a college student going through college with study loans), and her older brothers who work in Nairobi. Nevertheless, she does not mind being regarded as someone who is well off.

Part of the Community

George admits that owning a mobile phone makes some think you have made it big, although he does not agree with the assumption. As he sees it anyone can own a mobile phone, for example, you could get one as a present, just like he got his mobile phone from a friend. This might sound like a politically correct view, but at the same time George is very honest with the inner feeling that owning a mobile phone creates. With a mobile phone he is part of the community, and not left out like he sometimes felt before, when he had to borrow his partner's phone, or use the phone booth.

> 'I feel like I am like everyone else because I have a phone. Before I used to feel that I've been left out or something. Now I am part of the community.'
>
> **George, 26, Nairobi**

64% of those surveyed in Mumbai, 85% in Tanzania, 66% in Mexico and 79% in South Africa said they had more contact and better relationships with family and friends as a result of mobile phones (Ericsson Consumer Lab research).

MARKET DEVELOPMENT

The task of converting poor consumer into a user of advanced mobile services and application is part of the market development. The market development involves both the consumer as well as operators serving them. We will consider the risks and benefits to operators little bit later.

Here we shall reflect on the incentives and the benefits for the consumer, who is far isolated from the access to information society.

Market development is about understanding what consumers want and then creating the capacity to consume. The traditional approach to creating capacity to consume has been to provide the product or service cheap or free of charge. This has the feel of philanthropy. Charity might be good but it rarely solves the problem in a sustainable fashion. A rapidly evolving approach to encouraging consumption and choice is to make unit packages that are small and hence affordable. Pre-paid therefore has been a phenomenal success. Lower the domination of the pre-paid faster the adoption.

Market development is about reaching out to consumers and by that we mean *making sure we know what consumers think and want, and act upon it.*

As telecom sector professionals and key stakeholders, our ideas of the perspective of consumers on communications pricing and tariff reform may be based on any or all of the following misguided myths:

- That price is the most important factor for consumers, and they want to pay as little as possible;
- Mobile phones cost too much for the poor;
- That consumers take a short term view, and don't care about what impact it can make on the quality of their lives;
- That the poor cannot afford mobile tariffs, are not willing to pay them, and must always be offered very cheap services, even if this means heavy subsidies and lower levels of service;
- That providing tariff and handset subsidies is the best and most effective way to help the poor;
- That we already know more or less what consumers want, can afford and are willing to pay for;
- Mobile phones follow wealth; after a country becomes richer, it's people can afford more mobile phones;
- One should aid the consumers at the bottom of the pyramid and not make profit from them;
- Mobile communications serve the secondary needs of the poor rural consumers;
- That high tech mobile services and applications are meant for high-end subscribers and are too complicated to explain to consumers at bottom of the pyramid.

None of the Above is True!

Experience shows that consumers, including the poor, may have many more concerns that just the price of telecoms, and may be willing to pay for more reliable service, applications that are relevant for them, better quality or greater access.

We tried to examine each of these myths of consumer perspective in turn, and then have discussed method of outreach to bring consumers into the dialogue to help create what they want. Emergence of user-generated content (UGC) concepts e.g. *YouTube* clearly demonstrate the power of consumer participation in development of services and products suitable for them.

Myth #1: Mobile Services Cost Too Much for the Poor

The fact is that the cost of mobile services is falling faster than ever. As the tariffs go down the uptake goes up. The operators are driving profitability through economies of scale. Let us look at India, for instance, where mobile voice and data tariffs are some of the lowest in the world and still service providers on average are achieving EBITDA levels above 30%. There are about 4 billion of the world's population that sit below an average income of less than US$5 a day. The only way we can make the mobile services affordable is to remove all the hurdles to uptake and then offer tariffs that are bearable. The combination of these two acts are already proven to drive mass markets. Using this philosophy India is aiming to hit penetration levels of close to 50%.

Myth #2: That the Poor Cannot Afford Mobile Tariffs, Are Not Willing to Pay Them, and Must Always be Offered Very Cheap Services, Even if This Means Heavy Subsidies and Lower Levels of Service

The fact is that the poor often pay a higher price for making calls. If we compare the tariff structures, the lower you go on the prepaid denomination, the higher the price per minute of use. If we analyse this, we quickly realise that the poorest of poor who can only afford to buy the lowest denominations are paying premium per minute of use. On the other hand the high contract subscriber gets the best possible tariff per minute plus subsidies on handsets or in many cases free latest models of mobile phones. This is what I call the *reverse Robin Hood phenomena* – the poorest of poor funding the subsidies for rich. However, analyse this from another perspective, and it proves the point that there is a premium to earn if you as a telecommunications company serve the consumers at the bottom of the pyramid. The unserved and unmet consumers, the majority of whom live in far flung rural areas, represent a premium that if exploited can provide a future growth for telecom operators and service providers. In turn these consumers get access to communication and information.

Myth #3: One Should Aid the Poor, Not Profit From Them

The fact is that serving mobile communication to the poor offers additional benefit. The use of mobile services improves economy, quality of life and well being. Mobile phones are profitable and thus prove their usefulness, irrespective of whether they are used by the rich or the poor. To illustrate the point let us look at sub-Saharan Africa. In the last 5 to 7 years the industry has grown more than 100 fold. All the investment has come from private sector with solid business case; return on investment propositions. Aid money has been funding some research, mostly proving how difficult it is to do business in the region. However, we can say with confidence that very little investment has been directed towards actual development of communications infrastructure. As previously mentioned, the average consumer in developing economies like nations in sub-Saharan Africa is paying more than twice the average price per minute in comparison to consumers in developed economies.

Myth #4: Mobile Phones Serve Secondary Needs. One Needs to Focus on the Primary Needs of the Poor

The fact is that if the poor are empowered or enriched through mobile communication they can assert their own needs and better meet their primary needs. I believe that mobile communication is one basic human need. Use of mobile phones has proven to have a positive impact on the economy of the user. Our research in number of developing countries has proven time and again that mobile phone not only have positive impact on individual and national economy but also make a positive impact on the quality of life of individuals, improves their ability to create social networks and have access to information as and when they want it. Mobile phones (enabling communication), are thus not only a basic human need but they help to catalyse growth in meeting the other needs.

Myth #5: Mobile Phones Follow Wealth. After a Country Becomes Richer, It's People Can Afford Access to Mobile Phones

The fact is that wealth follows mobile phones. Ericsson studies in a number of developing nations such as Nigeria, Tanzania, Uganda, India and Indonesia have shown that each dollar spent on mobile infrastructure had a positive impact on the economy. It was also found that the lower the GDP, the higher the impact of mobile penetration on the economy. For example, every dollar spent on mobile communication in Nigeria gives approximately US$3 to US$4 to the Nigerian economy. The same statistics holds true for other fast-developing nations. The deployment of mobile phone networks is helping to make nations richer.

Problem of Measurement

Communication services are highly valued in every culture and at every income level and, in the last five years, developing economies have seen rapid growth in mobile telephony. In developing countries, the average amount spent on telecommunications is much higher than that in developed economies. A recent study conducted by Ericsson revealed that people with little money are often willing to spend between 10 to 20% of their earnings into mobile communications services. However, the same phenomenon of small businesses providing services such as plumbing or taxi services relying solely on a mobile phone exists in the developed world, indicating that the mobility inherent in mobile phones is itself valued. Historically, low rates of penetration in developing economies, coupled with rates of uptake that have exceeded expectations, should have made technology providers and investors eager to enter these markets, but until now operators have cited low Average Revenues per User (ARPU), long payback periods on investments, and murky regulatory environments as reasons for avoiding developing country markets. In addition to these challenges on the supply side of the market, the demand side also differs from that found in the developed world.

In looking at demand for mobile telephony in developing economies, subscriber numbers do not tell the whole story, since in rural areas one mobile phone is often used by many people. To this one can add the problem of flawed income data in countries where large sections of the economy exist solely in the informal sector. Income statistics for developing economies, especially at the lower end of the consumer pyramid, have proved to be a very unreliable predictor of mobile usage because of informal markets and the unreliability of existing income data.

Usage Patterns Vary in Different Markets

In addition to being characterised by unprecedented uptake in most of the developing economies, the introduction of mobile services has brought about a change in the business and operating climate, competing mobile operators have helped create an environment that fosters innovation at a ground breaking level.

In profound ways, demand for communications services in developing economies differs from demand patterns in developed countries. An individual consumer is much more likely to share a phone as opposed to owning it solely. Often, a mobile phone is considered a family asset as opposed to a personal asset. Moreover, a lack of legacy infrastructure extends beyond telephony. Due to a lack of financial infrastructure, including bank credit cards, automated teller machines (ATMs) etc. 50 to 70 percent of transactions today in developing economies are cash-based. Money that stays in the informal sector results in missed economic development opportunities, diminished opportunities for financial intermediation and a reduction in the efficiency of monetary policy. In somewhat paradoxical fashion, a lack of legacy infrastructure serves to reinforce demand for mobile telephony, as opportunity costs associated with physical travel can be so much higher. In the absence of roads and transportation, information and communication services are even more highly valued, as is demonstrated by examples through out this book.

Rural and Urban Consumers are Different

For the residents of rural communities, mobile communication services have great potential for positive economic and social impacts, definitely much higher than for their urban counterparts, since alternatives are few. Mobile communication services reduce travel needs, assist in job hunting, and provide better access to business information. These benefits occur even among the poorest communities. Unfortunately, assessing demand for information and communication technologies among the rural poor in developing economies is even more difficult than it is for the urban poor. Difficulties relating to poor roads and language differences hinder the process of surveying demand among this constituency.

Predicting Demand for Communication Services

Understanding what consumers at the bottom of the pyramid need and want and creating products and services that satisfy these needs and wants. *'If we build it, they will come,'* was the famous reprise from the movie

Field of Dreams[8]. It has since served to codify a philosophy of service intro-duction and to represent the core of an old argument. Do services have to be known and quantified before an investment in infrastructure, or will revenue-producing services materialise when society is enabled with suit-able infrastructure? Telecom engineers often invoke the Field of Dreams scenario, while accountants demand the business cases and marketers demand the market analyses. In traditional models, telecom engineers were dominating the decision power, hence go to market was based on inside-out philosophy. It is paramount that the approach is turned upside-down, it could be very expensive exercise to use that approach while ser-vicing bottom of the pyramid. The approach recommended henceforth is outside-in. Understanding what the users want, and customising the tech-nology to map these wants.

Prahalad[9] outlines several important considerations when marketing to poorer customers. Many of Prahalad's suggestions focus on tailoring the product or service to the local environment. Local environment refers not only to the actual product or service itself, but also to language localisa-tion and constraints imposed by intermittent or non-existent power sources, for example. Prahalad also stresses the importance of scalability and scope. Since 'bottom of the pyramid' (BOP) markets are large, solu-tions that are developed must be scalable, and developers must focus on the whole 'platform', designing products or services that can easily incor-porate new features as circumstances warrant. Additionally, developers must be prepared to educate consumers on usage, in effect creating demand. Above all, developers must focus on the price performance of their products at every step in the value chain: Process innovations are just as critical in BOP markets as product innovations.

THE NEED FOR NEW PRODUCTS AND SERVICES

When we review the most important breakthroughs in mobile communi-cations during the last decade, most of them have had little to do with technology. Perhaps the single innovation that has made the mobile com-munications the world's most penetrating service is the invention of prepaid tariffs – little things that made a really big difference.

It lends perspective to mobile communications to realise that the major-ity of the people in the world have never made a telephone call. The fact that a large percentage of the population in many countries is very young

[8] http://en.wikipedia.org/wiki/Field_of_Dreams
[9] C. K. Prahlad; *The Fortune at the Bottom of the Pyramid: Eradicating Poverty through Profits*. Wharton Publishing School.

partially accounts for this startling observation. However, it is sobering to observe that the mobile and fixed telephone density (percentage with telephones) in many emerging nations hovers around 5–10%. While many of us in technology focus on advanced data networking, the majority of the world has much more simple needs. I have on occasion held the momentary belief that a mobile phone would offer little improvement in the quality of life of a starving person in a rural area of an emerging nation. But one is disabused of this notion. For example, it has been observed that poor rural farmers raise their standard of living after getting access to mobile phone, for the simple reason that they are then able to find new customers and to discover for perhaps the first time the true market value of their products.

Mobile technology offers the possibility of creating instant infrastructure for the unwired world. Bringing the mobile phone to the world at large is proving to be a great feat in itself, but now of course we have the Internet overhanging over mobile telephony. At the current growth rate, Internet usage over mobile is expected to exceed internet usage over fixed line. Conceivably, all new users in the developing economies will be connected to the Internet using mobile technologies like WCDMA/HSPA. They could then, of course, use mobile Internet for voice telephony, as will many users in developed countries in this time frame (5–7 years).

In fact, the paradigm may shift from today's Internet-over-mobile to tomorrow's Mobile-over-Internet. Tomorrow's successful communications services will bridge the gap between work and home, and will forge a support web for business nomads.

Telecommuting has not spread as fast as has been predicted, and it is not necessarily good that everyone is in continuous communications with their work environment. In our homes many of us will have broadband communications access and home local area networks.

Services that aid us in e-Commerce, that bring us entertainment, that save us time and that create small conveniences will be among those that are and will be appreciated even by the consumers in the rural areas.

Philosophy for Developing Products and Services

The users from bottom of the pyramid will challenge the dominant logic of mobile service providers (the beliefs and the values that services providers serving the developed markets have been socialised with). For example, the basic economics of the low end mobile user are based on the small unit packages, low margin per unit, high volume and high return on capital employed. This is different from large unit packs, high margin per unit, high volume and reasonable return on capital employed. This shift in business economics is well exploited by offering lowest possible denomination of the prepaid cards lower the value of prepaid top-up card, higher the value and margin.

Key Principles of Innovation

The key principles of innovating products and services for low end of consumer pyramid are:

- *Focus on price performance of products and services*
  Serving the low end of consumer pyramid is not just about lower prices. It is about creating a new price-performance envelope.
- *Innovation requires local customisation*
  Low end consumer needs and aspirations cannot be solved with traditional services and applications. Service providers need to find local content that is relevant for their tastes and aspirations.
- *The BOP market is massive, content and services that are developed must be customised and localised*
  Content and services need to be mapped through the castes, cultures and languages. How does one take a service and product from the south of India to the north of Africa? From South America, to China, to India? Services and content must be designed in such a way that it is easy to adapt these to similar BOP markets. This is a key consideration for gaining volume and deriving economies of scale.
- *Product development must start from a deep understanding of the consumer needs*
  Marginal changes to products developed for the rich consumers in Europe, the USA and Japan will not do.
- *Deskilling work is critical*
  Most users in the low end consumer segment are poor. Many of them can't read and write. Hence, ease of use is of critical importance when creating products and services for low end users, for example: from SMS to Voice SMS, English SMS to Local Language SMS.
- *Education of users on product usage is the key*
  Innovations in educating the semiliterate customer segment on the use of new products can pose interesting challenges. Further, most of these users live in *media absent* areas, meaning they don't have access to TV media.

BEYOND VOICE APPLICATIONS – ENRICHED COMMUNICATION

Consumers in rural areas have repeatedly proven to be technology savvy, opposing the myth that consumers at the bottom of the pyramid can't use the data services since they can't read and write and hence, don't have ability to use the service.

The SMS service is truly the first so-called data service that has experienced the mass market. The reasons for that are simple, there were absolutely no hurdles to uptake, all handsets were capable of sending and receiving SMSs, no registration fee, no subscription fee and it is free to receive. This is fundamental to make a service mass market in the low end of the consumer pyramid. Service producers have to ensure that they have removed all the barriers to uptake. Figure 4.2, below, illustrates the ease of use of SMS services.

How shall service providers ensure a mass market appeal for mobile Internet access? Following are the key hurdles to making it a mass market service:

1) Cost of the handset
2) Battery life of the smart phone or computer
3) Cost of battery back up
4) To use smart phone the consumer needs proper training

In order to get access to such a service a consumer needs either a *smart phone* or a computer. Both smart phones and computers are very

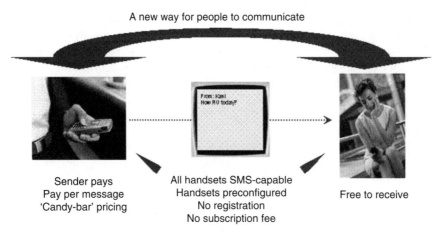

A new way for people to communicate

Sender pays	All handsets SMS-capable	
Pay per message	Handsets preconfigured	Free to receive
'Candy-bar' pricing	No registration	
	No subscription fee	

Figure 4.2 SMS has been a huge success due to there being no barriers to uptake

expensive. However if a service provider offers an innovative pricing model, for example the service provider subsidizes the terminal or finances it for a period and charges a portion on every top-up card the consumer buys, then it is affordable for the consumer. The back up battery can be part of the initial product package. The consumer is trained to use the terminal for application relevant to the user.

Developing Economies – Leaders

The Philippines are the m-payment pioneers, which is unsurprising since the country's 30 million mobile users send 620 billion text messages each month. Filipinos can already cash electronic cheques using mobile phone based cash remittance services. Credit is transferred from one person's phone to another's, and the recipient then nominates a sum of money to transfer to a commercial organisation's mobile account and finally withdraws cash at the nearest outlet for a small fee. Commercial partners include businesses with mass outlets, such as McDonalds, so it is quick and easy for users to access their money.

Peer-to-peer (P2P) top-ups for prepaid phones, where friends and family can transfer mobile credit to each other, are another popular m-payment application. Indeed, this has the most realistic potential of widespread uptake across the region, specifically involving very small, micro, amounts of credit.

Micro M-Payment

Known as micro m-payments, this is a niche area that is booming in developing economies in the South East, a key driver being the economic situation of the majority of the local population. While low income is not a barrier to mobile phone penetration in developing economies, where mobile phone penetration reached over 50% by the end of 2006, the vast majority of developing countries' mobile users have very low salaries and it is likely that they only have the need, and means, to transfer very small amounts of money.

This unique, local need represents a major revenue-boosting opportunity for operators in the region. Physical top-up vouchers, which have traditionally been used to top-up credit, are costly to produce, preventing operators from offering their subscribers the ultra-low denominations they require since they need to be able to recoup production costs. Introducing voucher-less top-up systems such as P2P top-ups, will cut costs and demonstrate significant returns, allowing operators to let customers top-up amounts as desired. Denominations of as little as 5 cents encourage usage amongst even the lowest-income population.

Operators can also reduce costs further by delivering P2P top-ups via the Unstructured Supplementary Data (USSD) channel. Although less well known than GPRS, USSD capability is available on all GSM handsets and does not require costly SIM upgrades. Therefore, the potential for reaching the pre-installed user base in developing economies in the South East is seen as its greatest asset.

Local Demand Drives Innovation

Despite progression in m-payments, the south eastern developing economies mobile phone markets are not generally more advanced than their European counterparts. The key differentiator in the case of m-payments is the alternative demands of the local market.

Micro m-payments, P2P or otherwise, are already a success because users only want to purchase small amounts at a time. Similarly, there is more scope for more advanced m-payment options to become popular in developing economies (compared with Europe, for example) because of the comparatively small limited number of people who have bank accounts.

As the majority of the population does not have the tools (bank accounts, credit cards, debit cards) to conduct electronic transactions as people in the West might recognise them, there is an opportunity for local operators to develop mobile-payment options. Developing economies clearly want to use mobile phones and, as the banking infrastructure is not clearly defined, there are fewer hurdles to overcome to convince users of the benefits of using a handset as a debit card or cash transfer tool.

Compare this with Europe, where the tools are already deeply embedded within society. Yes, the technology is available and relatively straightforward to deploy, but there exist legal and cultural challenges in Europe that suggest widespread uptake of m-payments will not happen.

Countries with developing economies have another significant advantage in that they operate in a considerably more liberal regulatory environment, meaning that m-payment opportunities are more easily exploited. Stringent EU guidelines about the transfer of money are a roadblock to widespread implementation on the scale that is happening in developing economies.

Moving Forward

The evidence suggests that the concept of micro m-payments has already been accepted in the Philippines and if Filipino operators continue to make this a success it will only be a matter of time before other south eastern developing countries follow suit. And moving beyond micro m-payments, operators may already be looking towards enabling users to buy physical goods with their mobile phones. If the popularity of micro m-payments in the Philippines is a barometer for success, operators in developing economies and beyond will be watching.

Case Study 1: *Mobile phones and financial transactions*

Mobile network operators are continuously introducing new products and services, to meet the ever growing demand for mobility in today's fast moving world, while simultaneously attempting to make such services increasingly relevant to, and affordable by the poorest members of society. One innovative application of mobile technology, that has attracted considerable attention in recent months, is the provision of financial services.

In many developing countries, particularly in rural areas, access to financial services is very limited resulting in a large percentage of the population operating on a cash basis only and outside of the formal banking system. However, the proliferation of mobile services in these countries has created a unique opportunity to provide financial services over the mobile network. In light of the growing size of international and national remittances, this opportunity could have significant implications.

In the developing economies however, there is a very large 'underclass' that is totally reliant on cash for all their day-to-day expenses. Moreover, this underclass makes no use of the banking sector and so is 'invisible' in terms of its cash value. At the same time, the need for cash forces the providers of goods and services in these markets to have adequate cash-handling facilities and this comes at some cost.

In these cases, the commercial organisations have much more to gain by addressing the problem of cash transactions. Not only is the risk associated with cash holdings that much greater, but the time-value of the cash being held outside the banking sector is entirely lost. Furthermore, the population in this category is lost, i.e. unseen by the banking sector. For these reasons, there is likely to be more incentive in developing economies to move the population at large away from cash, than exists in developed economies. That being so, a solution that meets the needs of developing economies will also have extensive application in the developed economies. This arises because the solution must be accompanied by very low costs as, if it were otherwise, the solution would have no appeal in those developing economies. The resulting low cost solutions can then be applied in the developed economies resulting in further efficiency gains.

THE FORGOTTEN RURAL POOR USERS

In developed economies it can be demonstrated that the population has wide and easy access to the banking system, with an extensive uptake of EFTPOS[10] and credit cards. Virtually all retail establishments have facilities for accepting both types of transaction in addition to cash and such facilities can often be found in service areas such as taxi cabs and even parking meters. However, there are many economies where such wide acceptance of a 'cashless' society is many years away. These are primarily economies where the average income is low with many people having no involvement with a bank at any time in their life. These people survive on cash and they very likely have no trust that a bank would serve their interests very well. At the same time, many banks would regard this segment of the market as being unprofitable and likely to be more costly to service. In many cases, the cash assets held by any one individual would be too small for any bank to regard as having value when considered alongside the traditional costs to maintain banking records for a customer. With the advent of mobile, that situation has changed in a subtle way but the change has not yet been recognised in many markets. Specifically, the same market segment that has shunned the banks and the associated electronic funds transfer systems, has contributed to the very high growth of prepaid mobile services in these markets. These users are often characterised by the need to communicate but without the complexity of a formal account with the network operator. They are invariably given prepaid service and while some markets insist on knowing the identity of the user, at least at the time the connection is activated, there is no certainty that this information is accurate in the long term. This class of user prefers anonymity and that is what they can often get with prepaid mobile. Topping up their account is as simple as buying a new prepaid card from a retailer and entering the details into the phone's keypad. Given that there are now large numbers of prepaid users in developing markets who are very familiar with using their phones for text and voice messaging as well as refilling their credit balance on the prepaid system, this same group is an ideal segment to target with a micro-payment feature. In many cases they have no relationship with any bank, do not use EFTPOS or credit cards and yet they have the ability to perform financial transactions as evidenced by their ability to purchase and activate prepaid cards for additional credit.

This potential has been recognised in a few markets, with probably the most success in the Philippines.

---

[10] EFTPOS: electronic funds transfer at point of sale.

The Significance of This Segment

The advantage of developing a market for micropayments or m-Commerce is that it continues to drive the economic system toward a cashless transaction environment. Elimination or minimisation of physical cash has many advantages including less opportunity for fraudulent or criminal activity, reduction of cash handling costs and, for the user, less reliance on having the right amount of cash when needed. It also allows the value of money to be better utilised. Cash held outside the banking system is not available for short-term investment so that the time-value of the cash asset is lost.

In the more affluent economies, there is already a good infrastructure for a cashless environment with most people having bank accounts and an array of both debit and credit cards. Nevertheless there is an underlying need for cash for minor purchases but there is little incentive to eliminate cash entirely. These economies can manage quite well and there is no specific interest group that feels sufficiently under pressure to develop systems aimed at eliminating cash from the environment. Systems that have been developed in such markets are often expensive and hence not particularly attractive to the user.

The Impact of Mobile Services

When GLOBE Telecom introduced GSM mobile services in the early 1990s, it attacked the high revenue end of the market with an emphasis on post-paid services, but with the later development of prepaid technology, the company commenced a prepaid GSM service in 2000. When SMART eventually commenced GSM services in 1999, it had little option but to seriously address the traditional prepaid segment of the market, which it did believing that profitability could come from a low margin, high volume business.

Following its introduction of service, SMART in the Philippines researched the uptake and the way in which the customers used the service and found that their lowest value prepaid recharge card was still too expensive for many of the users. These users could not afford the minimum P300 (US$6) charge for the card. As a result, the minimum card value was reduced to P30 or US60¢. This new value was quickly recognised by the market, resulting in a very high customer growth rate for SMART. This experience reinforced the view that the lower groups in the economic pyramid are influenced not only by the price of the product but also by the cost of the smallest element of that product, i.e. the segment was conforming to its characteristic of 'sachet' purchasing. Following SMART's necessary efforts to cater to the lower socio-economic classes to secure a customer base, it became apparent to the company that there was a definite future in this largely untapped market segment provided costs and hence the service charges could be kept at a very low level. The lesson from the prepaid recharge card value alone was an incentive to further reduce the recharge values but that could only be done if there was a move away from the more usual scratch cards to an electronic 'over-the-air' (OTA) system. As a result, the company turned its attention to technologies that would allow such a move and recognised that such technologies could deliver much greater value than just prepaid recharge using OTA concepts, and that gave rise to SMART's launch of their first mobile banking and commerce service, SMART Money, in December 2000.

Since then, various changes to the product have been made and it has been joined in the market place by GLOBE Telecom's own mobile money remittance and payment service called G-Cash. This new GLOBE service resulted from the company's desire to develop a unique m-commerce solution and was eventually launched in 2004.

Today in the Philippines, the emphasis in the mobile markets is on low transaction costs (text messaging at typically US2¢) and minimal re-charge values (US60¢) coupled with ease of use and a range of transactional applications aimed at addressing the population's needs. Both GLOBE's G-Cash and SMART's SMART Money are being actively promoted and from the information supplied, both are experiencing a high uptake of the services.

These same low 'sachet' costs have significantly influenced overall mobile demand, which now exceeds 35 million users out of a population variously, estimated at between 85 and 90 million. This mobile penetration level presents both major operators with enormous potential in the area of m-commerce.

SMART MONEY

SMART Money[11] is the product offered by SMART Communications. It has a related product, SMART Mobile Banking that allows customers to transfer funds from their bank accounts to their SMART mobile service account including prepaid recharging. These two products can coexist with some customers.

SMART Money was first introduced by SMART in December 2000 and has gone through several iterations to the present time. As currently configured, the service appears to address most requirements for a good and reliable micropayment platform.

[11] SmartMoney; Smart Communications Inc. Corporate Website http://www.smart.com.ph/SMART;, accessed on 24 October 2006.

The SMART Money product is essentially a facility for linking the user's phone to a 'cash' account.

Facilities exist for the customer to deposit cash, withdraw cash, top up the mobile phone prepaid credit levels from that account or other bank accounts, all without going near a bank or a SMART office if need be. Inherent in the operation is an ability to transfer credit between mobile users, so as well as allowing a semi-formal cash transfer, it allows the users to manipulate their credit in the system to suit their particular needs. It operates entirely on a credit basis, i.e. funds must be in the system before a customer can manipulate those funds. As a result bad debt is not an issue. The system gives no credit to the users. In its simplest form, SMART Money has no outwardly obvious attributes. The service is menu-driven from the phone and the customer can perform all necessary actions using the phone alone. In its preferred form, the service is coupled to a bank debit card as can be found in any community around the world. In this case the customer is provided with a card issued under the Master-Card banner, that can be used anywhere a normal debit card can be used, i.e. in ATMs, shops etc. The service is aimed at providing a wide range of transaction capabilities all of which should have considerable appeal to the target market. It is coupled into an account held by Banco de Oro (BDO) so that the user is effectively operating a BDO account using the phone as the transaction medium.

The specific list of features provided by SMART Money is as follows:

- Cash deposits
- Cash withdrawals
- Transfers of credit to the prepaid account (re-charge of prepaid service – SMARTLoad)
- Transfers of cash to and from other users
- Transfers of airtime credit from one user to another (SMART Pasa Load)
- Cashless purchasing at a wide range of shops where the retailer has a SMART Money account
- Cashless purchasing at any MasterCard-enabled retailer with a MasterCard debit card
- Direct credit from employer payroll
- Bill payment
- Inward international remittances from Overseas Filipino Workers (SMART Padala).

Customer Sign-up: For the customer wishing to become a SMART Money customer, the application process is relatively simple.

The normal situation involves the customer visiting a SMART office and signing up for the service. This may involve a SIM card change and it is necessary for the customer to have the phone available at that time.

A cash deposit into the account is not required at that time but without a cash balance in the account, no purchases or withdrawals can be made. However, having opened the account, others can transfer credit into it (see below). The customer will normally be encouraged to sign up for the associated debit card at a cost of P220 which includes the first year's subscription charge. Subsequent years attract the same charge, i.e. P220 per annum. Assuming the card is uplifted by the customer, it is prepared while the customer is being signed up and given to the customer at that time. If the customer applies at an office that has no card-printing facility, the card can be mailed or picked up over the next day or two. If the customer chooses to not have a card, then there is no charge for the sign-up. Under normal circumstances, the customer will make a cash deposit into the system at this point. As an added method of subscription, the company has provided an over-the-air activation process which registers the customer on SMART Money and allows credit transfers to the account but until the customer visits a SMART office and provides the necessary ID as required in the Philippines, cash withdrawals and purchases are not possible. This feature is specifically aimed at the SMART Padala feature described below.

Automatic Transaction Update

This feature, which is built into all the transaction services, provides for a text message to be sent to the user whenever a transaction is performed, whether by the use of the phone directly or by way of the bank debit card if that option has been chosen. This feature operates at all times on all transactions and provides the customer with a level of confidence in the use of the product. As an added safeguard, the customer has free access to the current credit balance using the menu on the phone and can also request a printed statement of the transactions at a nominal charge.

Cash-In

This feature allows the user to deposit cash into the user's account. As it is a physical medium, the cash must be deposited at one of the designated cash deposit locations. This includes SMART and BDO offices along with a range of accredited retailers who have agreed to take deposits. Every deposit must be covered by an acceptable ID in accordance with the requirements of the country's central bank (BSP). If the customer does not have the available debit card then the deposit is manual and the depositor must fill in a deposit form that requires a formal ID. If the depositor has opted for the debit card, then that can be used in a cash deposit terminal available at some locations. The terminal accepts the card and currency notes in payment with a minimum note size of P100 and the minimum deposit is set at P500. Note that these are country-specific requirements unrelated to the technology. As noted above, as soon as the cash has been deposited, the customer will receive a text message in addition to a paper receipt from the cash teller or automatic teller.

Cash-Out

This is the reverse of the cash-in procedure but is potentially a more flexible arrangement. In this case the customer can withdraw from a bank or SMART cashier or accredited retailer in exactly the same way as depositing cash. Again, a withdrawal must be completed with acceptable ID. However, users who have opted for the additional bank debit card can use that card in ATMs worldwide that accept MasterCard transactions; in that case the card provides the ID link. Subject to retailer policy, cash withdrawal from retail establishments using the debit card is technically possible although it is not a practice in the Philippines market.

Retail Purchasing

In this transaction, the user has two options. One is to use the debit card, in which case the purchase is done according to normal practice where a debit card is used. The other alternative can be used at participating retailers, and that involves the retailer originating the transaction request through his own SMART mobile phone terminal. Subject to the customer having a credit balance to cover the intended purchase, the customer receives an authorisation request via SMS. Once authorisation is given, the retailer and customer accounts are updated and the customer receives a confirmation of the transaction via SMS.

Credit Transfers

These are convenience transactions for the customer, allowing the transfer of a credit balance to another customer. The customer initiates a text message indicating the amount of the transfer and the SMART Money customer to whom the transfer is directed. Subject to fund availability, both parties receive a confirmation SMS.

An extension of this service is marketed by SMART under the banner of SMART Padala and is aimed at the significant number of migrant workers from the Philippines who regularly transfer funds to family back in the Philippines. For this, SMART, in conjunction with BDO, have established links with TRAVELEX, a worldwide group specialising in money transfers and foreign exchange cash conversions with outlets at airports and in cities in many countries. In the case of these Forex transfers, the migrant worker from the Philippines must know the mobile number of the Filipino family member and that family member must have a SMART Money account. Recognising that initially at least, many SMART customers will not also be signed on for SMART Money, the company has provided an over-the-air registration method for such users to quickly register for SMART Money. While this will quickly register the customer for the purposes of an international transfer, the customer must still visit a SMART or BDO office to collect the cash. The same concept allows employers to make direct credits to customers' SMART Money accounts.

Airtime Transfers

This is another form of credit transfer where the SMART Money customer can transfer airtime credit to another SMART Mobile customer's mobile account. In reality, this transfer can be done without being a SMART Money customer as there is no cash transaction with all credit records being held within the SMART prepaid system. The service is marketed as SMART Pasa Load.

Prepaid Top-Up

This is a natural extension of the service, using either SMART Money or their mobile banking feature, SMART Mobile Banking. It allows transfers of credit from a SMART Money or bank account direct to the prepaid card account. The minimum top-up from a cash account is P30 (US57¢).

Merchant Opportunities

As an adjunct to prepaid recharging, SMART Money offers participating retailers the opportunity to sell airtime to customers in lieu of previously used prepaid scratch cards. It is marketed by SMART as SMART Load. As long as the retailers have credit in their SMART accounts, they can sell units of airtime to prepaid users who do not have the capability to do so themselves. The smallest transfer recognised by SMART is a value of P2 (US4¢).

RELEVANCE OF MOBILE DATA AND HIGH BANDWIDTH SERVICES FOR THE RURAL / LOW END USER

The broader assumption is that mobile data and high bandwidth services are meant for high-end and high-brow users. Rural low-end users first of all can't afford these services, secondly don't find use for them and thirdly, rural consumers are mostly illiterate and cannot use these services.

Well to clarify this, we have tried to look some key mobile data and high bandwidth services and how they are being used by high-end or western users and than we tried to capture the possible relevance of these services to low-end users.

It is correct that low-end / rural users can't afford the same tariffs as the high-end western users. Secondly, the penetration of smart mobile phones and personal computers is very low; hence even though we see the relevance of these services to the rural low-end users, they don't have the possibility of using these services. However, if these hurdles are removed somehow, we could see obvious mass market appeal of these services in the rural low-end segment. What is very important is that the mobile data and high bandwidth services need to tailored to the local needs and wants. Also it must be realised that most of these users live in far-flung rural areas where they still lack supporting infrastructures like electricity, local content providers etc. Mobile service providers need to come up with innovative business and operations models that first remove all the hurdles to uptake; secondly they need to create local content that is directly relevant for the rural users, offers a business benefit, enhances the quality of work and justifies the return on investment.

The big consumer brands have ambitions to penetrate the low-end rural consumer segment. The only problem they have is how to reach to these users from advertising and branding standpoint. Banks and financial institutions also have ambitions to enter this segment in order to achieve the mass market for their services. *Grameen Banking Model* has paved the road and so are many other industries plus the donor communities, Western international development agencies, World Bank etc. If all these industries cross pollenate and work together I believe there is a win-win.

The Grameen model emerged from the poor-focused grassroots institution, the Grameen Bank, started by Professor Mohammed Yunus in Bangladesh. It essentially adopts the following methodology. A bank unit is set up with a Field Manager and a number of bank workers, covering an area of about 15 to 22 villages. The manager and workers start by visiting villages to familiarise themeselves with the local milieu in which they will be operating and identify prospective clientele; they also explain the purpose, functions and mode of operation of the bank to the local population. Groups of five prospective borrowers are formed: in the first stage, only two of them are eligible for, and receive, a loan. The group is observed for a month to see if the members are conforming to rules of the bank. Only if the first two borrowers repay the principal plus interest over a period of fifty weeks do other members of the group become eligible themselves for a loan. Because of these restrictions, there is substantial group pressure to keep individual records clear. In this sense, collective responsibility of the group serves as collateral on the loan.[12]

Wireless broadband connects a home or business to the Internet using a radio link between the customer's location and the service provider's facility. Wireless broadband can be mobile or fixed. Wireless technologies using longer range directional equipment provide broadband service in remote or sparsely populated areas where DSL or cable modem service would be costly to provide. Speeds are generally comparable to DSL and cable modem. An external antenna is usually required. Fixed wireless broadband service is becoming more and more widely available at airports, city parks, bookstores, and other public locations called 'hotspots.' Hotspots generally use a short-range technology that provides speeds up to 54 Mbps. Wireless fidelity (Wi-Fi) technology is also often used in conjunction with DSL or cable modem service to connect devices within a home or business to the Internet via a broadband connection.

Mobile wireless broadband services are also becoming available from mobile telephone service providers and others. These services are generally appropriate for highly mobile customers and require a special PC card with a built in antenna that plugs into a user's laptop computer. With the evolution of mobile technologies like GSM, the uplink and down link speeds are becomes faster and faster. With HSPA speeds up to 15MB per second are becoming possible.

Mobile broadband therefore provides technical capability to access a wide range of resources, services, and products that can enhance their life in a variety of ways. These resources, services, and products include, but are not limited to:

- **Education, culture and entertainment**
  Broadband can overcome geographical and financial barriers to provide access to a wide range of educational, cultural, and recreational opportunities and resources.
- **Telehealth and telemedicine**
  Mobille broadband can facilitate provision of medical care to unserved and underserved populations through remote diagnosis, treatment, monitoring, and consultations with specialists.

---

[12] http://www.grameen-info.org/bank/mcredit/cmodel.html Last accessed June 2007.

- **Economic development and e-commerce**
  Broadband can promote economic development and revitalisation through electronic commerce by creating new jobs and attracting new industries and providing access to regional, national, and worldwide markets.
- **Electronic government (e-government)**
  Electronic government can help streamline people's interaction with government agencies, and provide information about government policies, procedures, benefits, and programmes.
- **Public safety and homeland security**
  Broadband can help protect the public by facilitating and promoting public safety information and procedures, including, but not limited to: 1) early warning/public alert systems and disaster preparation programmes 2) remote security monitoring and real time security background checks 3) backup systems for public safety communications networks.
- **Broadband communications services**
  Broadband provides access to new telecommunications technologies such as Voice Over Internet Protocol (VoIP) allowing voice communication using the Internet.
- **Communications services for people with disabilities**
  Broadband permits users of Telecommunications Relay Services (TRS) to use Video Relay Services (VRS) to communicate more easily, quickly and expressively with voice telephone users.

Let us look at some popular mobile data and high bandwidth services that are being offered by mobile services providers across the globe.

Mobile Radio

This is an audio streaming service that gives consumer possibility to listen to different FM radio stations. This is been tried in different parts of the world has proven to be very successful. This is very relevant for the low-end user.

User Location Asset Tracking

This geographical positioning system can be used to track your assets. Recently the service has been deployed to track cattle. People in the West are using this service to track their pets. Transportation companies are using this service to track their trucks. In the USA, for example, Verizon combined forces with MobileGates to create FuelFinder, a location-based service that helps users find the best gas prices in their area. When requested, FuelFinder prompts users to enter a zip code or city, and within seconds, the service displays national high, low, and average fuel prices, along with an interactive map with plotted results. When the user selects a gas station, they are given detailed, turn-by-turn directions. FuelFinder has over 110 000 gas stations from coast to coast in its database, which is updated every 2 hours. FuelFinder is available exclusively to Verizon Mobile Web 2.0 subscribers.

Video Streaming

Video streaming allows users to check the traffic condition of various roads, streamed live on their handset anytime, anywhere. This service provides users with video streaming of live road traffic. Video streaming traffic monitoring can only be accessed on PDA or computer.

Navigation Services

Navigation services offer the ability to navigate from one location to another, using audio or visual directions. This service may also include the use of maps.

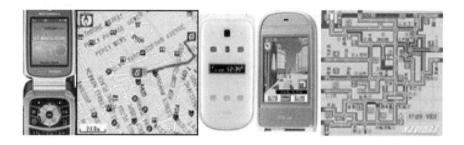

For example, with Optus Zoo FindA users can get detailed maps and directions in the palm of their hand. FindA offers several types of location based services. FindA includes the following valuable position location services: 1) FindMe: Provides a detailed map of where the user is located and allows them to zoom out to orientate themselves. They can browse in each direction around their location so they get a clear view of where they are. 2) FindPlace: Provides a detailed map of any location. The user can specify a place name, suburb or address to get the map they want. They can zoom in and out and browse in each direction to get a clear view of their location. 3) FindNearby: Provides a list of pharmacies, shops, restaurants, cafes, etc. that are closest to the user. They can select one and receive the address, phone number and a detailed map of the place they've chosen and additionally opt to get directions from where they are. 4) Directions: Provides detailed step-by-step driving directions from one spot to another, on easy-to-use maps, with helpful landmarks and pointers to look out for on the way. The maps provide details on freeways, intersections and streets, so the user doesn't get lost on their way to somewhere new. The maps also show how far away a destination is so it's easy for the user to plan their trip. It even has a quick reverse link to help the user find their way back home again.

Internet Access

Offers full Internet connectivity to a computing device (e.g., PC notebook, desktop, Internet appliance, etc.) using a 3G PC card, embedded module, router, or a phone as a modem.

This service helps consumers to access Internet. It is useful to give information access to consumers, particularly those living in rural far-flung areas with no other means of accessing the world-wide-web and being part true information society.

Video Mail (Video Messaging)

For example MTN's new VideoMail service changes the way users communicate with family, friends and business associates. VideoMail functions as a communications centre for video calls. The service works just like VoiceMail, but takes it one step further. When they aren't available to accept a video call, VideoMail users can greet callers face-to-face and invite them to leave a video message. Subscribers can retrieve their VideoMail by either making a video call to their video mailbox, have the message delivered to their 3G mobile handset via MMS, or view it on their PC via e-mail.

Advertising Surveys

Offers the ability to view selective advertising on a 3G mobile device. This service may include gift certificates, coupons, and advertising based on position location (Proximity). For instance Virgin Mobile USA (VMU) customers that subscribe to the Sugar Mama service can earn free airtime minutes by providing feedback to VMU on short advertisements. Customers that subscribe to Sugar Mama offer VMU a minute of their time, and receive a minute of airtime in return. Sugar Mama works in three different ways: 1) ADTIME: watch short online video spots and provide VMU with feedback 2) TEXTIME: answer questions via text messages 3) QTIME: fill out surveys regarding brands and their products and services. Sugar Mama lets users earn up to 75 minutes of airtime per month.

Mobile Music

Offers the ability to download and play user selected music or audio programmes. Users can replay the audio files at their discretion.

The mobile music service allows users to download sound tracks, photographs and ringtones on their mobile phone. Users are able to browse through a WAP listing of available artists and genres before selecting the songs they wish to download. Value-added services are also provided such as extended content related to artists and bands, such as wallpaper and ring tones. Users are provided with their selected songs on both their mobile handset and their PC. The service is very relevant as long as the music content is customised to local choice, local records etc.

Telemedicine

Offers the ability to remotely monitor patients and provide medial services with the help of 3G wireless devices and networks. This is shown in figure 4.3, below.

For instance, KTF (part of Korean Telecom) OK Phone Doctor service allows users to alleviate common headaches and pains through audio-visual therapy played through their mobile device The OK Phone Doctor service offers users the ability to listen to chosen music frequencies and color wavelengths known to be of medical benefit to sooth and calm injury. Users simply download and install the program onto their mobile device and update the additional content as required (new audio tracks, etc.). The system plays a programme for different symptoms, repeating every 1 to 2 minutes – the user simply needs to focus on the content to alleviate stress and headaches.

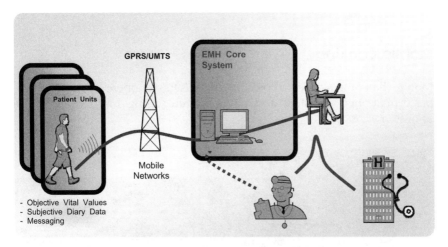

Figure 4.3 Mobile technology deployed as health application

Homewatch

This is particularly useful in area where security is an issue or for a working couple who have their small children taken care of by a helper or babysitter.

Homewatch service allows users to make visual contact with their home and office via an installed web camera. This service can be used for monitoring activities at home or workplace. The number of Homewatch points depends on user's choice. This is pretty relevant service, in rural context the user actually have to physically travel miles to check the home or asset he has left behind.

Mobile Banking

Mobile banking service allows mobile users to open bank accounts, perform money transfers and payments etc. using his or her mobile phone.

For instance, Far Eastone Indonesia offers I-style mobile banking. I-style synchronises the mobile phone with various banking activities such as transactions, payments etc. This service enables the user to access ATM function of the banks. This service allows the user to make transfers, payments and inquiries through the handset. Participating banks include Far Eastern International Bank and United World Chinese Commercial Bank.

Mobile TV

Mobile TV service allows terrestrial transmission of TV programmes and multimedia services to mobile phones via radio frequencies.

For instance Reliance Mobile TV in India allows users to watch 'live' television or download hourly video clips using their 3G mobile phones. Reliance Mobile TV users can download live video from the following Indian news channels: TIMES NOW, AajTak, IndiaTV, CNBC, NDTV 24 × 7. Reliance Mobile TV users can stream live video from the following online Indian news channels: AajTak Online, NDTV Online.

Instant Messaging

Offers the ability to instantaneously exchange text messages between two or more people using mobile devices. This is also known as 'chatting'. The service may also alert the user as to who in their 'buddy list' is online (presence) and allow them to exchange messages within large groups (chat rooms). Some common protocols include AOL, Yahoo!, MSN or Jabber.

For instance, 3 MSN Messenger service from Microsoft allows users on the 3 network to chat and communicate in real-time via instant messages with other users on 3 phones and PCs using the MSN protocol. Messages

sent to other users are sent in real-time and a user can see the response and reply immediately. The service allows users to sign-in to their MSN account with a Microsoft Passport, as they would do on their PC. The user is able to access their friends and groups listings, recent messages and change their online status. With 3 Yahoo!, users can check their emails, send instant messages, upload photos, search the web, and synchronise their address book with Yahoo! Go on the Nokia N73. The range of Yahoo! Services from search to photos to ringtones and graphics is accessible either via the Internet browser on the handset or through client applications such as Yahoo! Ready or Yahoo! Go for Mobile on select devices.

Video Messaging

Offers the ability to capture, send and receive short video clips. The service may also include online movie albums.

VideoSMS allows users to send and receive multi-media messages, enhanced with both motion and sound. In addition to text, VideoSMS customers can also communicate their message visually. VideoSMS brings messaging to life and allows users to share spectacular sights and special moments with their loved ones. First, users take videos using their 3G camera enabled mobile phone. Then, they can share the moment by

sending it to family and friends. The multimedia message can be viewed using an MMS compatible phone which is equipped with a video player. If the receiving user does not have a video playing capable phone, they can view the message via Vodacom's online portal at Vodacom4me.

Mobile Transactions

Offers the ability to store financial account and personal authentication information on a 3G mobile device and use the information to perform mobile payments, banking transactions, stock and bond trades, etc.

For instance, Reliance in India offers TravelnShop which makes traveling easy, by allowing subscribers to make arrangements for their travel through their 3G mobile phones. The following are services provided by

TravelnShop: check availability of train tickets and book online; get information on city bus routes; get information on airline schedules, along with fares and book airline travel online; discover places to visit and learn how to get there, where to eat and where to shop.

Another example is EZ Felica offered by KDDI, Japan. EZ Felica is a mobile commerce service that allows subscribers to make e-payments, pick-up tickets, board airlines, enter their homes and even sign-in at karaoke bars, all with a tap of their phone handsets. EZ Felica enabled phones can directly download and store monetary value (prepaid money), display of account balances and usage history, authenticate a person's identification or membership, purchase products at a point of sale, execute electronic ticketing, and collect online credit and loyalty points/awards.

Games Download

Offers the ability to download games to a 3G mobile device over the 3G network. With this service, users can play games on mobile with full colour, sound and vibration effects.

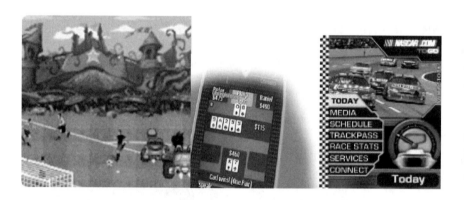

Larger 3G games contain better imagery, sound quality and offer a longer-lasting gaming experience. 3G download games include some multiplayer aspects.

For instance, AT&T offer Cingular Games. Users can download games with full-colour animation and robust sound effects. Once downloaded, users can play the games again and again, creating a little diversion during a busy day. Cingular Games users can choose from the following game genres: 3D Games, Action & Adventure, Strategy and Role Playing Games Sports, Card, Board & Casino, Gameroom, Star Wars, HBO, Disney and Cartoon Network, Puzzle, Retro, Play for Prizes.

Web Browsing and E-Mails

Mobile web browsing offers the ability to browse web pages on the screen of a mobile device (e.g., WAP, iMode, HTML, etc) and e-mail offers the ability to send and receive e-mails

Google in your hand.

GO TO MARKET APPROACH – KEY PRINCIPLES

When approaching BOP market, the lower segment of the consumer pyramid, we ensure we follow the following key principles:

Be First to Enter This Segment

This market segment is generally very loyal. Hence approaching this segment proactively has long lasting advantage. We can analyse this by looking at the existing established consumer brands in the BOP market and impact of the ones that approached this market first. Below is the list of some of the established brands and consumer loyalty towards these. One of the key reasons why these brands replace the names of the products was that these were first to enter the BOP market. For instance tea referred as Lipton, cornflakes as Kellogg's, cold drink as Coke, shaving blades as Gillette, washing powder as Surf, mouthwash as Listerine etc.

Develop Products Using Outside-In Approach

The traditional approach has been to bring the products and packages that did well in high-end market, remove all frills and fancies and then offer it the BOP. This was a push technique. Many times this failed, particularly when offering data and multimedia services. The right approach is to understand the needs and wants of these users, feed the input to the product development team of the service providers, tweak the technology, customise the content and map the product offering to the exact needs of the users, for example, for the users that can't read and write, so that they will be able to leave video or voice message instead of SMS and e-mail.

Convenience and Ease of Use

The product or service must make life more efficient, is simple in functionality e.g. similar user interfaces for most services (and across accesses), communications and content services available between different device types (e.g. video call from 3G phone to PC), services adapted to the device and access characteristics, simple processes for identification and payments (e.g. single sign on), should easy to have cost control.

Always Best Connected

Able to connect anywhere, anytime, by device of choice, when on the move. Best varies according to user preferences, the service used, quality, speed and price. Seamless transition between access methods for voice/video/packet data (session continuity).

Ensure Reliability and Security

Reliability in all transactions independent of access, connection quality (voice, video, data). Authentication and security must be reliable regardless of access. No viruses, no worms, no fraud Integrity -nobody listening in and ability to see who is calling.

Services Linked to Consumer Needs and Wants

Services must innovate based on what is relevant for the users, application can be simple and down to earth but mapped to needs. This shall drive volumes and spin-off impact, which shall lead to mass market appeal for the service or product.

Business Case Through Lower Tariffs High Growth

The tariffs of the products must not follow cost plus principle prevalent in traditional model. The prices and tariffs should be based on the affordability of the users. The business case and return on investment must be driven by low tariff and high volume principles. Low margin/high growth follows the economies of scale principle.

CONCLUSION

In conclusion, it has been found that the consumers at the bottom of the pyramid have aspirations and needs for using mobile voice, data and advanced high bandwidth services. Service providers must adapt their business models to address this segment, remove all the hurdles to uptake and offer affordable tariffs. The business case for service providers shall be driven through economies of scale; call it low ARPU/high growth. Service providers need to create partnerships with industry stakeholders, involve other industries in their game plan, influence governments and regulators to lower the tax burdens, engage donor communities etc. to ensure lower total cost of ownership and in the end create affordable tariffs for the end consumers. Advanced mobile data services are very relevant, in fact more relevant than they ever were for high-end western user who already had access to world-wide-web thorough fixed broadband line and other technologies. A mobile phone might be the first PC the user at the bottom of the pyramid has ever seen. Enabling it with high bandwidth technologies and making services and applications relevant to the user will truly bridge the so-called digital divide.

5

Mobilising Wireless Communications for Mass Markets

OVERVIEW

Providing communications to mass markets holds enormous promise for universal commercial growth. More than two billion people in the world do not have access to basics such as fridges, stoves, televisions, clothes, phones and so on, that can support a comfortable and healthy lifestyle. In this chapter, we look at mobilising the mass market by looking at what still needs to be done to include the mass market in the global marketplace and how this can be done. We try to establish who the main drivers are, or should be, in directing efforts toward mobilising the mass communications markets and we try to understand the important question: Who bells the cat? Who really drives communications development and growth? The case for rural telephony and the operator's business case are also presented. Donor communities also have an important role to play in mobilising communications for mass markets in developing countries and the effectiveness of donor communities is evaluated.

Communications technology developments have been explosive and some of the main technology developments such as global services mobile (GSM), code division multiple access (CDMA) and WiMax are briefly explored to determine what wireless technologies are available to operators. Here, we also look at some equipment developments such as micro base-stations, co-siting, refurbished communication equipment (reuse of old GSM equipment) that operators could use to assist in spreading communications to mass markets cost effectively. The implications for increasing teledensity diffusion, using low-cost handsets and SIM cards are also

Business Models for Sustainable Telecoms Growth in Developing Economies
S. Kaul, F. Ali, S. Janakiram and B. Wattenström
© 2008 S. Kaul, F. Ali, S. Janakiram and B. Wattenström

touched on. We conclude this chapter by discussing some mobile service developments such as mobile TV, messaging, infotainment, mobile Internet, video service and other innovations that include airtime transfer, increasing prepaid customer loyalty and how these can be applied to the mass market.

MOBILISING MASS MARKET ECONOMIES

There is a sharp skew in global income distribution. Some 80% of the world's wealth is concentrated in the hands of 20% of the worlds population (a very fortunate few) and more than a billion people today live on less than US$1 while another billion are making ends meet with just between US$1 and US$2 a day. This means that about 80% of world income, is consumed by a mere fifth of its population. Clearly there is serious imbalance in global income.

Global Income Segmentation

Table 5.1 shows the distribution of average annual global population income per quintile. The classification of quintiles spreads across the globe and includes developed and developing countries. Table 5.1 shows that average annual global income can be segmented into five main categories or quintiles with the highest quintile (top 20%) receiving on average US$25 000 annually and representing the 20% wealthiest across all nations, while the lowest quintile receives an average annual income of $3650 and represents a third of the world's population.[1]

Table 5.1 Average Annual Global Income per Quintile

	Average Income in 000's of $'s
Highest Quintile	25 000
Fourth Quintile	7 000
Third Quintile	3 000
Second Quintile	730
Lowest Quintile	365

Adapted from: Infoplease, 2007[2]

[1] Infoplease.   http://www.infoplease.com/cig/economics/worldwide-market. html last accessed on June 9, 2007.
[2] Infoplease.   http://www.infoplease.com/cig/economics/worldwide-market. html last accessed on June 9, 2007.

The poverty of the world's poorest has a major impact on global economic wellbeing in the near, medium and longer term and there are a number of serious moral implications that go straight through to our consciences. However, the focus of this chapter is not on the moral aspects of poverty The aim here is to highlight the repercussions that ongoing poverty has for the international marketplace.

There is a fundamental problem with creating more economic growth in the world today. Clearly, the developed markets appear to have matured and the industrialised nations of the world cannot sell goods and services to consumers that live on less than US$2 a day. The majority of the poor in mass markets spend their whole lives in poverty and most don't know what it is like to fly in an aeroplane, own a television, drive a motorcar or even use a mobile or fixed line phone. Given such a bleak situation, how is it possible to mobilise the mass market to buy communication products or services? How is it possible to sell communications or for that matter any type of goods and/or services to a market where there is no home ownership? How can we sell anything to a population that expend their total energy in simply getting basic needs satisfied? The answer is that it is impossible.

Looking at the third quintile in table 5.1, we note that income is larger than the lowest two quintiles, but even at this level most people are surviving without some of the basic comforts taken for granted in the developed world, such as an electric heater or refrigerator. At this level, average annual income per household is about US$3000 per year, which translates to just $8.20 a day. This is very small compared to the highest and fourth quintiles. The three lowest quintiles in table 5.1 make up about 3.7 billion people, or a third of the world's 6.2 billion population. Yet, they possess less than one third of the purchasing power of people in the top quintile. Undoubtedly, they constitute a huge potential market, but unfortunately they lack the level of income required to make them marketable. It is only in the fourth quintile that individuals generally earn around $19 per day. At this level, people have sufficient disposable income that extends beyond that required, meeting their needs for bare necessities.

Today, the majority of marketing initiatives are focused at reaching the wealthy segment of global population, meaning that marketing effort is directed at only 20% of global population and excludes about 80% of the true potential market that exists in the world.[3] This means that four fifths of the world's population are still underserviced. Prahalad and Hart (2002) argue the emerging-market strategies of the majority of multinational corporations (MNC's) throughout the last decade or so has concentrated on the 'wealthy few' or 'emerging middle-income earner' and not on the 'real source of market promise' which is the 'billions of aspiring

[3] Infoplease.   http://www.infoplease.com/cig/economics/worldwide-market. html last accessed on June 9, 2007.

poor who are joining the market economy for the first time.'[4] The need for developing countries to continue to grow their economies is real and therefore, there is a burning need for them to rethink their global economic strategies.

Need for New Communications Mass Market Mobilisation Strategy

Until now, marketing efforts have been directed at only 20% of global population. The majority of the world's population have been excluded from the mainstream economic activities that the developed world is so accustomed to. One such important economic activity is communications. Almost all of the world's rich have access to fixed and/or wireless communications while the majority of people in the world have none. This doesn't mean that 80% of world population do not need communications. In fact, all that it means is that only a fifth of people are able to afford communications. For us to really understand the economic potential that lies at the bottom of the pyramid Prahalad[5] identifies a number of conventional assumptions the developed world about has about mass market consumers in developing countries:

- Poor consumers are not targeted because current cost structures do not allow firms to compete in the market.
- Poor consumers do not have a need for products and services and cannot afford the same kind of products and services that are marketed in developed markets.
- Only consumers in developed markets are willing and able to pay for new technology functionalities.
- Poor consumers can be given access to older technology and be satisfied.
- Mass market has no significance to the business in the longer term.
- Bottom of the pyramid's needs may be satisfied by governments, NGOs and other non profitable organisations because managers in developed countries do not view developing country business environments, that are skewed in favour of humanitarian efforts, as viable.
- Talented management believe intellectual stimulation and excitement can only be found in the developed markets and therefore it is difficult to recruit the best managers to serve mass markets that lie at the bottom of the pyramid.

[4,5] Prahalad, C.K and Hart, L.S. 2002. The Fortune at the Bottom of the Pyramid. *Strategy + Business* 26, 1–14.

These assumptions ferment an already deceptive management paradigm and conceal the true value that resides in mass markets right at the bottom of the pyramid. In a sense it promulgates false management logic, very much like the case of the South African diamond prospector who found a diamond on the riverside of the Limpopo river. Our good judgement allures us to the reality that if the prospector really found a diamond on the riverside there would have been a flood of prospectors to the site to search for more diamonds. Quite like the case of the Limpopo diamond, the bottom of the mass-market opportunity overrides traditional management logic. Undoubtedly, because firms in developed economies have not been drawn to the developing markets in droves, does not justify our thinking as managers that no unexplored profitable opportunity presents there.

Given the enormous propensity of the mass market for consumption (because this group also has needs and wants) it is now time to put on a different lens to view the mass market. The movement of time doesn't change, only people do and today the face of the world has changed because there are some four billion people with marginal income that demand global economic inclusion. The solution is to create economic enablers to increase their means of income. This means a set of vibrant and living strategies is required to deal with new times.

New investment needs to be targeted at mass markets where billions of potential consumers can be taken out of their state of poverty and desperation. Investment in this segment of the global market will mean avoiding longer-term social and political disintegration especially if the divide between rich and poor countries continues to exist. Mobilising the mass market needs disruptive innovations in technology and the development and application of fresh thinking to create sustainable business models. New thinking is required around the way value is built for communications products and services.

Strategic Communications Development – Circle of Cooperation

Global, national and regional communications development strategies and actions in different developing countries and regions of the world will dictate whether the availability of communications applications brings about social and economic development.[6]

---

[6] IDRC. 2007. http://www.idrc.ca/en/ev-28872-201-1-DO_TOPIC.html last accessed on June 22, 2007.

As depicted in figure 5.1, for communications development strategies to be effective they must include all stakeholders that have an interest in developing the global information society (GIS). To develop sustainable business models for developing country communication development, requires cooperation, shared understanding and vision between all the different stakeholders such as business, government, NGOs, regulators, end users and others.

As mass markets in developing countries are mobilised, there is a strong need for them to create effective policies and processes to maximise the benefits and exercise control over the inherent risks associated with ICT development. Effectively, this translates to having well coordinated action that encompasses promoting economic empowerment in the mass market, satisfying mass market customer needs for technologies and services, promoting government objectives and satisfying the requirements of business, end users, NGOs, regulators and others. Hence, mass market communication mobilisation strategies need to account for the vested interest of the various stakeholders and create a unified overlap that leads to the fulfilment of satisfying multi-stakeholder interest.

ICT policies, education, competence development and other programmes that assess technologies are needed to support the creative use of ICT. Hence, producing and using the new innovative ICT applications will definitely be different between countries and will depend on each country's ICT policies, the technological competence of its people and level of competence development and finally on how each country assesses the different ICT technologies. Therefore, a pressing requirement for communications development strategies is the formation of resource coalitions and strategic partnerships, alliances and mutual cooperation between multiple stakeholders.

Figure 5.1   Circle of stakeholder cooperation

Crafting communications development strategies for communications development is not simple. There are no 'one size fits all' formulae for developing such strategies and no set of fixed rules to apply to the design of such strategies. Each developing country has unique needs and must follow distinct paths that hold relevance for them. Generally and at best it may be suggested that communications development strategies should set clear priorities that traverse across the fragmented borders of business sectors, government, government agencies, regulators, NGOs, customers and others with an interest. These should be followed through with well coordinated and clearly implemented targets and plans. These strategies must be flexible and open to the needs of the different stakeholder groups. Finally, these strategies including the social and technological building blocks have to be very flexible regarding what could or should be done. There should be clearly thought out directives supported with clear and simple action plans coupled to the right balance of funding and accountability.

Given the potential of ICT, all governments and other stakeholders need to build new capabilities to produce, access, and use these technologies. To build these capabilities, ICT strategies must be responsive to sustainable development goals and involve all social and economic stakeholders. Government occupies a central position in facilitating market development, promoting dialogue between stakeholders, initiating regulation that is effective, ensuring the promotion of cross sector stakeholder dialogue and ensuring the provision of public services that are relevant to country conditions.[7]

Communications Development Strategy Guidelines

The development of communications will offer tremendous economic and social benefits to all people in developing countries if the right blend of ICT strategies is implemented. Some of the main considerations to be taken into account when crafting and implementing ICT strategy should encompass the use of ICT for social and economic advancement, enhancing human resources development for effective strategy implementation, promoting innovation and access to communications networks to ensure sustainable long term development, establishing opportunities for investment financing and investment of communications, building end user product and service knowledge about communications services and benefits and implementing processes for influence and monitor international rules. Figure 5.2 shows the requisites for developing communications in developing markets.

[7] IDRC. 2007. http://www.idrc.ca/en/ev-28872-201-1-DO_TOPIC.html last accessed on June 22, 2007.

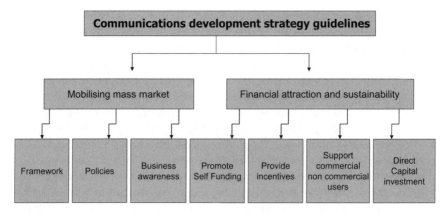

Figure 5.2 Communications development strategy guidelines in developing markets

As illustrated in figure 5.2, developing countries that want to build communications capability and develop a national communications infrastructure (NCI) will have to mobilise and pool large amounts of investment and expertise.[8] Mobilising mass markets for communications requires action in three main areas: (1) developing countries need to set out a framework, which establishes a market-friendly environment for investors to bring in foreign direct investment (FDI) (2) developing countries should implement policies that offer incentives and explicitly promote economic development for the poor especially in rural areas (3) business communities (especially MNCs) need to be made aware of the commercial possibilities and opportunities that reside in mass markets.

Financial attraction and sustainability for communications in developing countries can be achieved by (1) concentrating efforts towards promoting self-funded projects (2) promoting investment through incentives given to businesses that invest in rural communities for example providing tax incentives allowing investment deductions on capital investments over a period of time (3) supporting commercial and non-commercial users in remote areas (4) direct capital investment in partnership with business and others such as NGOs, subsidising ICT investment and integrating ICT into existing programmes and projects.

[8] IDRC. 2007. http://www.idrc.ca/en/ev-28872-201-1-DO_TOPIC.html last accessed on June 22, 2007.

MAIN DRIVERS – MASS MARKET COMMUNICATIONS MOBILISATION

All stakeholders have a critical part to play in mobilising the communications mass market. However, international bodies such as the World Trade Organisation (WTO), national regulators and MNCs have a greater role to play in global communications reform. This is mainly the case because these parties are powerful levers for reform. There is a strong reliance on the market's invisible hand in countries to foster growth in the ICT sector. However, the developing countries implementing communications reform are at quite diverse stages and therefore have different development priorities.[9] For example, there are different privatisation and competitive mixes in each country's communications industry. In some countries, the government still controls communications supply, while in others such as India, South Africa, Nigeria, Uganda, Tanzania, Indonesia, Malaysia, Russia, Botswana, governments have abandoned or are in the process of abandoning the monopolistic model in favour of allowing competitive entry into the sector.

World Trade Organisation (WTO)

The WTO has initiated the liberalisation process on basic telecommunication services by committing developing countries to opening up their communication markets to foreign suppliers and by influencing them to introduce competition in the market for the supply of all communications services. In this sense, at a global level the seed for mass communications growth has been sowed. Unfortunately some of the least developed countries have not signed the international WTO agreement and this will have serious repercussions for them in future, as they will need to develop the capabilities required for mass penetration of communications themselves.

Developing Country Commitment

On the other hand many developing countries have signed the accord and committed themselves to privatising and liberalising their communications markets, although many developing countries such as South Africa still continue to adopt a communications market communications

[9] IDRC. 2007. http://www.idrc.ca/en/ev-28872-201-1-DO_TOPIC.html last accessed on June 22, 2007.

structure where government continues to have a direct interest. Telkom SA the fixed line operator and 50% owner of Vodacom, for example, has been listed on the Johannesburg Stock Exchange (JSE), but is still partly owned by the South African government.

A direct interest in the communications market by government downplays the effectiveness of the free market system, because governments are almost always biased towards the companies they own and are covertly reluctant to take decisions that negatively affect their own interests in these companies. For example, in South Africa, there has been a huge calling by NGOs, business and other stakeholders to break the parastatal Telkom's hold over the local loop ('last mile') access network, because its charges for Internet and other value added services are perceived as being too high. Monopolising 'last mile' access prevents new entrants from competing on an equal footing with the incumbent. During 2007, rather than yield to external pressure, the state extended Telkom's monopoly over the 'last mile' by another four years. Such acts diminish the power of the invisible free market forces that are responsible for increasing communications penetration through open competition.

Operator and Service Provider Reluctance

Market liberalisation and pure competition bodes well for end users of communications because they generally lead to decreases in the prices charged for communication services. However, such developments also lead to major reductions in operator revenues. Hence, there is almost always strong reluctance on the side of communications operators to enter new markets where there are already established operators or markets that they perceive as unprofitable. Until now, operators entering developing markets have only set their sights on the more prosperous market segments where profits can be realised, totally ignoring the needs of the poorest because of their traditional belief that this segment offers no value for profit creation.

Operators have made little or no effort to develop the poor communications market segments in most developing countries. The poorest of the poor are not able to access communications, not because they do not have a need for the services, but mainly because they cannot afford these services and business has failed to, as Prahalad puts it, 'stop thinking of the poor as victims or as a burden and start recognising them as resilient and creative entrepreneurs and value-conscious consumers'.[10] An important question that arises then is who really bells the cat?

[10] Prahalad, C.K and Hart, L.S. 2002. The Fortune at the Bottom of the Pyramid. *Strategy + Business* 26, 1–14.

Who Bells the Cat? Who Drives Communication Development?

Until now, governments, NGOs and others have had to carry the burden for developing communications for mass markets. Communications operators in the form of MNCs and big business (operators, service providers and equipment manufacturers), have neglected poor market segments as having no purchasing power, but have failed to realise that these market segments can be transformed into the fastest-growing world market through innovations and new functionalities that are affordable and tailored to meet the needs of the poor. Cost cutting communications innovations do not just lead to more profits, they also serve to provide the poor in mass markets with affordable, quality products and connectivity to the world. The generally held belief that only governments and NGOs can reduce poverty and are responsible for meeting the communications needs of the poor stands on weak foundations, because the private sector has a very important role to play by providing communications cheaply and promoting the entrepreneurial talent of the poor.

Figure 5.4 shows how economically empowering poor communities in developing country mass markets sows the seeds to create capacity for the marginalised. As depicted in figure 5.4, empowering mass communications markets requires cooperation on a number of levels. Firstly, there

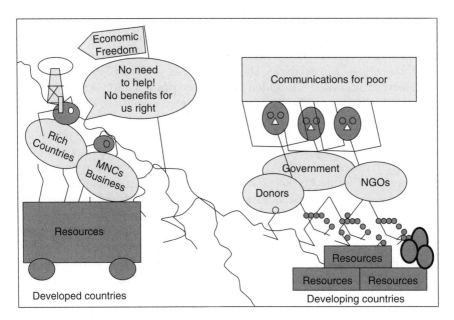

Figure 5.3   Who bells the cat? Who drives communication development?

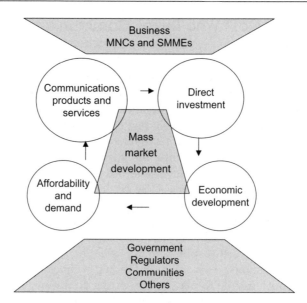

Figure 5.4 Economically empowering mass markets

is the need for business, both big and small, to work together as partners. Large business enterprises such as the MNCs need to make direct investments by providing small, medium and micro enterprises with direct financing. For example, financing the development of base stations and phone shops in rural areas.

Economic development and empowerment takes place when small and/or micro entrepreneurs are granted franchises for selling communications services, for example. These franchisees, in turn, empower poor people by providing value added access to communication services at low prices, which are used by locals to access a key ingredient in wealth creation: information. For example, a poor person in a rural community that wants to supply farm produce to the urban market will be able to use the community shop to phone and obtain information about prices, contacts and supply processes without incurring the costs associated with personally travelling to the market in the city. Such services enable the poor to rise above their circumstances and achieve financial independence by participating, at low costs, directly in the global economic arena.

Financial independence leads to an increase in the demand for communications products and/or services and so the cycle of growth begins. Government and regulators play an important role in setting up the framework for communications through their policies, procedures and legislation. Government may also partner with regulators in public private partnership (PPP) projects to ensure that communications development is channelled towards mass market development. NGOs, donor communities and others provide support for the establishment of communications development by directing funding to communications infrastructure development projects in developing areas, for example, in rural areas where communications development is seriously lacking.

Case study 1: *Safaricom empowering Kenyan farmers through communications*

Safaricom, a Vodafone subsidiary, has introduced mobile text messaging services that allow Kenyan farmers fast and easy access to agricultural market updates. By using the Sokoni Short Messaging Service (SMS) both buyers and sellers of agricultural commodities are able to keep in touch in real time with changing commodity prices via Safaricom mobile phones. In this way, end users are able to use the information to their advantage by securing favourable prices for their products. Reports from the Kenyan Agricultural Commodity Exchange (KACE) are beamed daily across the Safaricom network. End users of this service send an SMS name of the commodity they want information on for example 'coffee', 'maize', 'banana' and shortly after receive an updated commodities price at the market. Sokoni is a powerful business tool that makes it possible for farmers to access the most promising time and place for executing buy and sell decisions. KACE is a good example of an NGO that assists farmers and market traders to have a mutually good business relationship by providing invaluable daily reports about the daily price for commodities across all Kenyan markets. In this way traders are able to secure critical market information in real time daily instead of having to delay until when they can find the pricing in the next days newspapers. Sokoni enables traders to market their produce or even place bids for other goods, as well as posting short messages or questions they have about agriculture.

Toward Communications Affordability – Network Infrastructure Developments

We believe there a number of ways to improve affordability that can be applied to mass market communications development models (1) operators and service providers should make payments easy and affordable through instalments in the sale of cell phones, for example, where end users require this (2) operators and service providers must implement major cost cutting initiatives; being a lean, no frills operator reduces costs and makes way for further call price reduction benefits that can be passed on to end users (3) reduce the communications product and/or services such as communications airtime packages into smaller more easily affordable quantities (4) allow users to merely use products and services rather than insist on ownership; for example, provide pay on a by-use basis, such as selling prepaid airtime or community phones and internet use at an Internet café instead of customers buying a mobile phone or computer (5) simplify distribution by removing the inefficient layers created by the middleman, for example airtime top-ups; (6) work with other sectors/industries, the banking sector for example, to unlock value in mass markets, to provide services to the poor; (7) employ the right blend of technology to provide great services at affordable prices.

When determining the pricing for communications services, two main financial drivers influence operators and service provider's decision making. These are operational expenditure (OPEX) and capital expenditure (CAPEX). So it makes sense that when opex and capex are high, prices charged for services are also set high. Hence, if operators and service providers want to reduce costs they will need to reduce OPEX and CAPEX.

OPEX can be reduced through a number of significant cost cutting initiatives such as introducing more streamlined processes, redefining organisational structures and implementing information technology applications. CAPEX reduction, on the other hand, is a little more complicated. Providing adequate services to meet the needs of end users demands the right spread of technology equipment, for example: the number of radio base stations (RBS) deployed in a specific geographic location determines the coverage and quality of services (QoS) that the operator provides to its customers. As a result, because of the huge revenues generated from affluent customers, service providers and operators have no hesitation when injecting large sums into capital equipment. When it comes to investing in developing areas, there is a strong reluctance on their side because (1) the huge capital costs involved in procuring the required equipment and (2) it takes much longer to recoup these expenses from poorer customers because of their needs for more basic services and the prices that they are willing and able to pay is less than the more wealthy customers. However true this may have been in the past, today there have been a number of new technological developments that have made it possible for operators to minimise CAPEX and provide profitable communications services to the poor.

SOME KEY WIRELESS TECHNOLOGY DEVELOPMENTS

These are exciting and dynamic times for the communications world. Fixed, mobile and wireless technologies are speedily being adopted across every sector and are rapidly spreading across the workplace, home and among communities. Technological innovations advance productivity, make life simpler and enhance convenience. Due to the slow spread of fixed line developments, the adoption of fixed line as a single solution for increasing communications penetration in mass markets is not a feasible option for operators and service providers. The capital costs associated with developing a fixed line network are huge. Wireless technologies have most certainly been way ahead in providing solutions for mass market communications. Therefore, we have not considered fixed line technology developments, although there are many. We have opted instead to focus on wireless and mobile technology developments.

Widening the Net

As shown in figure 5.5, Ericsson's expander solution minimises time, risk and investment to capture the full potential of mass markets and simultaneously assists operators to meet the needs of existing and new users. The expander solution enables operators to achieve this by making full use of the radio network infrastructure (RNI). At the same time, the expander is able to provision for various empowering services that lead to solid operator revenue increases. The expander enables operators to grow network coverage cost effectively. This has resulted in substantial reductions in wasteful duplication of base stations and a subsequent reduction in operator CAPEX.

In this way, communications growth can be encouraged, because lower costs enable operators to offer services over a much larger geographic

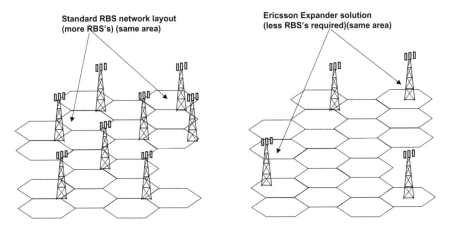

Figure 5.5 Ericsson RBS expander solution

spread at lower prices. Such innovations enable operators to benefit from the huge opportunity presented by the mass market while satisfying mass market end user needs.

Co-Siting and Location Services

Figure 5.6 shows a typical co-siting network infrastructure layout for a mobile operator. Co-siting is the term used to describe when different technology platforms share a common site. Co-siting enables the operator to reduce costs associated with renting and duplicating of communications facilities.

Today, sharing different technology platforms is possible in the core, transport and access layers of the communications network. For example, in figure 5.6 the core network is shared between the UMTS (3G) radio base station network controller and GSM base transmission station controller equipment. Similarly the antenna is shared by both GSM and WCDMA antenna equipment. Co-siting reduces capital costs associated with building of separate facilities to house equipment for different technologies. Such developments reduce operator capital expenditure. The rapid evolution in mobile technology is another important area that requires further

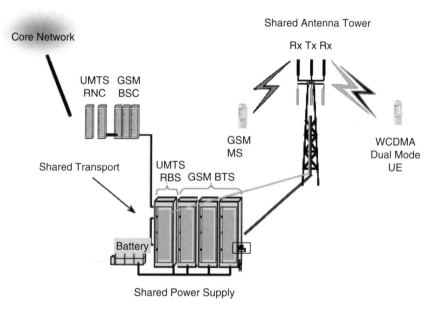

Figure 5.6 Typical simplified co-siting network infrastructure layout using GSM and WCDMA equipment

discussion because it has a direct bearing on communications penetration for developing markets.

Mobile Technology Evolution

As shown in figure 5.7, mobile technology developments have taken place along two main development pathways, using two main technology standards: GSM and code division multiplexing (CDMA). These developments have led to a number of evolved technologies in the GSM and CDMA standards band.

Since the mass adoption of wireless technology, there have been a number of stepped technological developments over time. The main drivers of these developments has been speed, coverage and capacity since these elements have a direct influence over the kinds of end user application services that can be offered at the service layer end of the open systems interface (OSI).

The kinds of end user services that communications operators and service providers are able to offer on a wireless network depends on the speed, coverage and amount of data that can travel over a spectrum of radio signals. Hence, wireless technology advancements have focused on increasing these important factors. The development path for GSM has been general packet radio switching (GPRS), enhanced data rates for GSM evolution (EDGE), third generation (3G) and more recently high-speed downlink packet access (HSDPA). CDMA has evolved along the path, CDMA 2000 1X and CDMA 2000 1X EVDO. GPRS, EDGE and CDMA 2000 1X were responsible for laying the foundations of third generation GSM

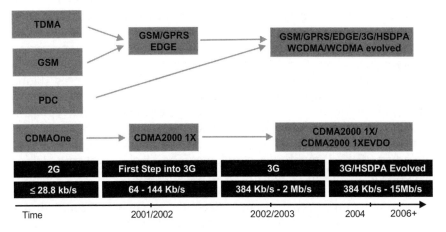

Figure 5.7 Mobile technology development pathways

and CDMA technology developments. These developments have resulted in the establishment of 3G, HSDPA for GSM and CDMA 2000 1X EVDO for CDMA.

Second generation (2G) mobile technologies, such as general packet radio switching (GPRS) and CDMA 2000 1X, have the capability to cover a wide area but at lower data throughput rates. Unfortunately, this imposes a serious limitation on the types of services that operators and service providers may provide to end users. 3G technologies on the other hand have the advantage of both increased coverage as well as increased throughput speed for data. This simply means that with 3G, operators and service providers can increase the bouquet of services (such as digital imaging, music and video) it offers customers.

The problem with 3G applications is still largely the limitation on speed, although coverage and QoS are good. Wireless local area network (WLAN) on the other hand enables operators to offer faster data rates than 3G, but coverage is limited. HSDPA and CDMA 2000 EVDO allows high area coverage and data throughput making it an ideal technology to offer a wide spread of great mobile services such as high speed Internet to a wider area.

To facilitate a high level understanding of mobile communications, the following sections briefly discuss each of the GSM, CDMA and WiMax technologies available to operators for mass-market communications infrastructure development.

GSM

GSM is a non-proprietary open technology platform that is in a state of constant evolution. A major advantage of GSM is its international roaming capability. With this capability, end users are able to enjoy standardised, seamless connectability on the same number in over than 170 countries. In places where terrestrial GSM services are not available, roaming via satellite enables users to still be in contact. GSM uses advanced digital technology and time division multiple access (TDMA) technology. With GSM, voice is digitally encoded using a specialised encoder that replicates speech characteristics. This method of transmission permits a very efficient data rate/information content ratio. High bandwidth services are already available through 2G technologies. The development path to third generation (3G) GSM (3GSM) is clearly mapped out and brings with it the possibilities of sophisticated data and multimedia applications. Expectations are that the GSM standard will continue to evolve through developments in wireless, satellite and cordless systems that will offer many advanced services such as high speed multimedia data services and support for simultaneous use of these services with seamless integration to the Internet and fixed line networks.[11]

General Packet Radio Service

General packet radio service (GPRS) is the world's most universal wireless data service that is available on almost every GSM network. Based on Internet Protocol (IP), GPRS mobile connectivity solutions support a wide array of enterprise and end user service applications. GPRS throughput rates extend up to 40 kilobits per second (Kbit/s) with the advantage of connecting to the global communications network from any place in the world. A major feature of GPRS is that end users can enjoy sophisticated and advanced data services such as mobile Internet browsing, mobile e-mail, video streaming, multimedia messaging and a variety of location-based services (LBS). Transmission speeds using GPRS is dependant on the type of mobile device that is used. Generally, most GPRS devices are able to download data at 24 Kbp/s although higher speeds of up to 171.2 Kbit/s are theoretically possible, when a single end user is assigned 8 slots. In reality 40–50 Kbps. PC cards capable of GPRS will send data up to speeds of 48 Kbps. Table 5.2 compares the speeds available for different fixed and mobile technologies.

Enhanced Data Rates for GSM Evolution

Further developments to GSM networks have been achieved with enhanced data rates for GSM Evolution (EDGE) technology. EDGE provides up to three times the data capacity of GPRS. Using EDGE, operators

Table 5.2   Fixed and mobile technology speeds – comparison

Technology	Uplink (Sending)	Downlink (Receiving)
GSM CSD	9.6–14 kbps	9.6–14 kbps
GPRS	14 kbps	28–64 kbps
Dial-UP	56 kbps	56 kbps
ISDN Standard	64 kbps	64 kbps
ADSL	256 kbps	1024 kbps
3G, HSDPA, CDMA 1XEVDO	Up to 2 Mbps	Up to 2 Mbps

Adapted from: Global Mobile Supplier Association (GMSA), 2007.[12]

[11] GSM World. 2007. http://www.gsmworld.org last accessed on June 27, 2007.
[12] Global Mobile Suppliers Association. 2007. www.gmsacom.com last accessed on July 4, 2007.

can handle three times more subscribers than GPRS; triple their data rate per subscriber, or add extra capacity to their voice communications. EDGE uses the same TDMA (Time Division Multiple Access) frame structure, logic channel and 200 kHz carrier bandwidth as today's GSM networks, which allows it to be overlaid directly onto an existing GSM network. For many existing GSM/GPRS networks, EDGE is a simple software-upgrade.

EDGE allows the delivery of advanced mobile services such as the downloading of video and music clips, full multimedia messaging, high-speed colour Internet access and e-mail on the move. Due to the very small incremental cost of including EDGE capability in GSM network deployment, virtually all new GSM infrastructure deployments are also EDGE capable and nearly all new mid- to high-level GSM devices also include EDGE radio technology. In November 2006, 156 commercial GSM/EDGE networks were in existence in 92 countries out of a total of 213 GSM/EDGE deployments in 118 countries.[13]

Third Generation GSM

Third generation GSM (3GSM) is a later addition to GSM. 3GSM, allows operators to provide a variety of mobile multimedia services, such as music, TV and video, rich entertainment content as well as Internet access. 3GSM services are delivered over a GSM network that has been modified using Wideband-CDMA (W-CDMA). Global operators and parties working together parties from the 3G Partnership Project (3GPP) standards organisation developed 3GSM as an open standard.

HSDPA

HSDPA, is the acronym for high-speed downlink packet access. HSDPA is a relatively new protocol for mobile telephone data transmission. HSDPA is a software upgrade for 3G that will deliver up to 5 times the data rates of standard 3GSM (W-CDMA). Of the 123 commercial 3GSM mobile networks live in 55 countries, more than half are HSDPA enabled today. In total, 121 networks across 55 countries have committed to deploying HSDPA.[14] It is known as a 3.5G technology. Essentially, the standard will provide download speeds on a mobile phone equivalent to an ADSL (Asymmetric Digital Subscriber Line) line in a home, removing

---

[13] Global Mobile Suppliers Association. 2007. www.gsacom.com last accessed on July 4, 2007.
[14] http://www.gsmworld.com/news/press_2006/press06_44.shtml

any limitations placed on the use of mobile phones by a slow connection. It is an evolution and improvement on W-CDMA. HSDPA improves the data transfer rate by a factor of at least five over W-CDMA. HSDPA can achieve theoretical data transmission speeds of 8–10 megabits per second (Mbps). Though any data can be transmitted, applications with high data demands such as video and streaming music are the focus of HSDPA.

HSDPA improves on W-CDMA by using different modulation and coding techniques. HSDPA establishes a new channel within W-CDMA called high-speed downlink shared channel (HS-DSCH). Using only downlink, HS-DSCH works differently to other channels and enables faster downlink speeds. Hence, data is sent directly from the source to the handset. With HSDPA, It data cannot be send from the mobile phone to a source. HS-DSCH enables the channel to be shared between multiple users that in turn allows the radio signals be effectively used for faster downloads.

The widespread availability of HSDPA may take a while to be realised, or it may never be achieved. Most countries did not have a widespread 3G network in place at the end of 2005. However, many mobile operators have been upgrading to deploy 3G networks that may be further upgraded to 3.5G at the time the market is ready for 3.5G services. Early bird HSDPA services will have speeds that range from 1.8 Mbps that will be upgraded to 3.6 Mbps as more advanced devices become available, that require faster speeds. HSDPA faces an uncertain future since it competes with other technology standards such as CDMA2000 1xEV-DO and WiMax that also provide high-speed capability.[15]

High Speed OFDM Packet Access

High Speed OFDM Packet Access (HSOPA) is a proposed part of 3GPP's Long Term Evolution (LTE) upgrade path for 3G systems. HSOPA is also often referred to as Super 3G. If adopted, HSOPA succeeds HSDPA and HSUPA technologies specified in 3GPP releases 5 and 6. Unlike HSDPA or HSUPA, HSOPA is an entirely new air interface system, unrelated to and incompatible with W-CDMAB.[16]

Open Loop Power Control

Figure 5.8 shows the way open loop RF power control is used in a mobile telephone system (MTS) to balance the level of radio frequency (RF)

[15] Wisegeek. 2007. http://www.wisegeek.com/what-is-hsdpa.htm
[16] Wilkipedia. 2007. http://en.wikipedia.org/wiki/HSOPA last accessed on July 5, 2007.

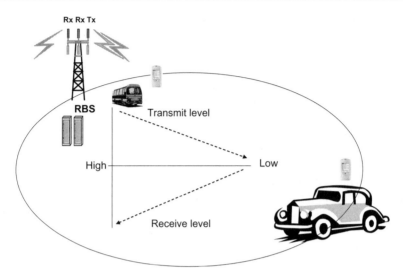

Figure 5.8 Open loop power control

signals that are received by the RBS from the mobile phone regardless of the distance between the mobile telephone and the RBS.

As illustrated in figure 5.8, the mobile phone's radio frequency (RF) amplifier adjustment is controlled by the feedback it receives from the receiver section. Throughout its time in the cell, the mobile phone continues to measure the strength of the radio signal that it receives from the RBS and calculates the signal strength loss between the base station and mobile phone. As the mobile telephone moves away from the base station, the radio signal received within the cell decreases. As signal received becomes stronger, the mobile telephone reduces its own RF signal output.

When the mobile signal received becomes weaker, the mobile phone amplifies its own RF signal. These results in the signal received at the RBS from the mobile phone remaining approximately at about the same power level, irrespective of the distance of the mobile telephone. At the same time it assists to save energy.

CDMA2000 System

Figure 5.9 provides a high level visual demonstration of a typical CDMA2000 radio system. As shown in figure 5.9, multiple devices can be used within the CDMA2000 system that includes CDMA2000, IS-95, multiple bandwidth radios and various CDMA compatible mobile phones.

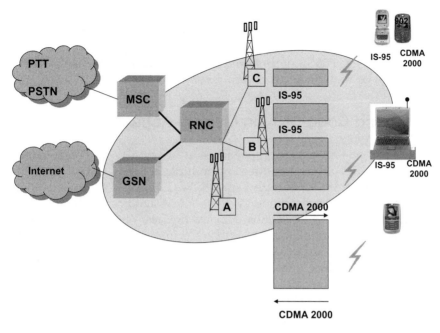

Figure 5.9 Typical CDMA 2000 system

Code Division Multiple Access (CDMA), is a mobile cellular technology originally known as IS-95. CDMA is a strong competitor to GSM technology. Today, there are many different variants of CDMA. The original CDMA is known as cdmaOne. The technology has evolved into cdma2000 with a number of variants such as 1X EV, 1XEV-DO, and MC 3X. The different variants make use of a 1.25 Mhz channel whereas; 3X uses a 5 Mhz channel. Developed by Qualcomm, and enhanced by Ericsson, Wideband CDMA (WCDMA) forms the cornerstones of 3G networks. CDMA is characterised by high capacity and small cell radius, employing spread-spectrum technology and a special coding scheme.[17]

Data capability has been improved by the 1XRTT CDMA standard that is an improvement over CDMAOne and enables operators to provide faster data throughput rates of up to 300 kbps. Capacity and increased battery life has been significant increased with 1XRTT CDMA. Today, like with GSM, global resources are being directed toward third-generation CDMA technology developments that include Multi-Carrier (CDMA2000 1xMC and HDR in 1.25 MHz bandwidth and 3xMC in 5 MHz bandwidth)

[17] Cellular. 2007. http://www.cellular.co.za/cdma.htm last accessed on July 3, 2007.

and Direct Spread (WCDMA in 5 MHz bandwidth). This first phase of CDMA2000 commonly referred to as 1XRTT, 3G1X, or simply 1X – was developed to double existing voice capacity and support always-on data transmission speeds 10 times faster than typically available today, some 153.6 kbps on both the forward and reverse links.[18]

Wideband Code-Division Multiple Access

Wideband code-division multiple access (W-CDMA) is an ITU standard that has evolved from code-division multiple access (CDMA). It is officially known as IMT-2000 direct spread. W-CDMA is a third-generation (3G) mobile wireless technology that offers significantly higher data speeds over mobile and portable wireless devices. W-CDMA technology supports a variety of high bandwidth voice, images, data and video communications at speeds of up to 2 Mbps for local area access or 384 Kbps for wide area access. Signals that are inputted are digitised, coded and transmitted using the spread-spectrum mode over a wide array of radio frequencies. WCDMA makes use of a 5 MHz-wide carrier, compared with narrowband CDMA that uses 200 KHz-wide carrier.[19]

Fast fact: NTT DoCoMo in Japan launched the world's first commercial W-CDMA (wideband-code division multiple access) service, FOMA, in 2001.

WiMax

Worldwide interoperability for microwave access (WiMax) is related to the IEEE 802.16 standard and its variants. The WiMax forum manages the evolution and relevance for market of WiMax that was originally designed to operate at 10–66 GHz. The WiMax system, however, had to change to offer broadband wireless access (BWA) in the 2–11 GHz frequency range. WiMax standard includes variants (profiles) that make use of different radio channel type combinations (single carrier –vs- multicarrier), modulation types, channel coding types to provide fixed, nomadic or portable services. The advantage of WiMax as a technology is that it can provide multiple types of services to the same user with different QoS levels. WiMax has been engineered to mix contention based (competitive access) and contention free (polled access) to provide services that have different quality of service (QoS) levels. A good example of this is the installation

[18] Cellular. 2007. http://www.cellular.co.za/cdma.htm last accessed on July 5, 2007.
[19] Wikipedia. 2007. http://wikipedia.org/wiki/w-cdma last accessed on June 7, 2007.

a single WiMax transceiver in an office building and provision of real time telephone services and best effort Internet browsing services on the same WiMax connection. WiMax protocols are designed to allow for point-to-point (PTP), point-to-multipoint (PMP) and mesh networks.

Another advantage of WiMax is that operators can use mesh configuration to link different base stations without the need to install or lease interconnecting communication lines substantially reducing opex. To enhance its appeal, WiMax operators can provide a number of additional communications services such as leased line, residential broadband, commercial broadband and digital television (IPTV) services. WiMax makes use of radio channel bandwidths that vary between 1.25 MHz to 28 MHz and it is possible to achieve data transmission rates that exceed 155 Mbps. WiMax data connections over WiMax radio channels include basic (physical connection), primary (device control), secondary (configuration) and transport (user data).

WiMax provides an excellent solution for extending the limited coverage of public WLAN (hotspots) to citywide coverage (hot zones) with the same technology being usable both in the at home environment as well as with mobility. WiMax is also very good for covering metropolitan areas for mobile data service delivery and can support fixed broadband access in both the urban and suburban areas where the quality of copper is generally poor, or where it is difficult to unbundle the local loop. Another advantage of WiMax is its ability to increase communications penetration in low-density areas where there are technical and/or economic factors that hinder broadband deployment. WiMax can additionally be positioned as a supporting solution to mobile that provides voice and constrained data services, through providing higher bandwidth as required, particularly in densely populated urban areas. Although public WLAN offers a number of benefits, it is constrained by its coverage and mobile limitations. WiMax overcomes these constraints and provides broadband connectivity in larger areas.

Broadband Wireless Access

Wireless solutions can complement fixed digital subscriber line (DSL) mainly in areas with a low density. Wireless solutions assist by connecting end users any time, anywhere and at appropriate speeds for broadband. One thing is clear: last-mile wireless technologies providing broadband connections to homes and businesses will have a major influence on the market, and wireless fidelity (WiFi) hotspots will change substantially as the demand for high-speed Internet increases in areas with low population densities. Mobile operators hoping to use their existing mobile infrastructure plan to develop new services that fixed operators do not provide like nomadic wireless DSL. Advances in microprocessors enable wireless systems to make provision for alternatives to DSL in those geographic areas where fixed line DSL would be near impossible to deploy due to geographic location such as the rural areas or because of economic cost reasons such as the high costs associated with installing copper fixed lines.[20]

Innovative technological developments have ensured that wireless systems are no longer fixed, but also provide nomadic access with the extended capability to allow mobility in the not so distant future. Although public WLAN offers high data rates (several Mbit/s) and is available in many places such as coffee shops, airports and hotels, it is severely constrained by its limited coverage and mobility capabilities that allow for public applications. The IEEE 802.16e variant of WiMax rises above these limitations allowing broadband connections to wider areas or so called hot zones. Today, WiMax enables broadband wireless to be accessed at very fast data rates of tens of Mbit/s over a broad range covering several kilometres.

Broadband wireless access (BWA) technology facilitates the following main applications that make it very useful for mass market mobilisation:

- *Wireless DSL:* offering DSL-like services that are wireless. The wireless nature of this technology enables it to be deployed at a much faster pace than wireline technology. This is mainly because there is no need to make any changes to the end users line.

- *Backhaul for WLAN:* Operators are increasingly being pressured to offer broadband services. To do this many operators are looking at using WiFi technology to provide broadband access to low population areas. Nonetheless, even if WiFi offers the ideal access solution, connecting to the point of presence (PoP) gateway will be problematic. BWA technology can be used effectively to overcome the constraints associated with this kind of problem.

- *Nomadic/mobile DSL:* This is a relatively new concept that has become possible because of new technological innovations. Mobile DSL makes use of non-line-of-sight, zero-install and plug-and-play features, making anytime, ubiquitous connections a reality. Both indoor and outdoor broadband Internet connections are possible with mobile DSL within the coverage area.

---

[20] Baldwin, P and Thomas, L. 2005. *Promoting Private Sector Investment and Innovation.To address the Information and Communication needs of the Poor in Sub Saharan Africa.* Report supervised Kerry McNamara and Seth Ayers from infoDev and Souheil Marine from Alcatel Digital Bridge Initiative department, InfoDev and Alcatel. Published by Alcatel.

Satellite Technology

Another enabling technology for mobilising mass market communications is satellite technology. Satellite communications networks are able to extend instant coverage to large under- and ill-equipped areas, or to supplement both fixed and mobile communications including other end user services such as TV and radio broadcasting, rapid Internet proliferation and content distribution by fixed or mobile multicast. Satellite technology provides global Internet and enables Internet service providers (ISPs) to establish ubiquitous global operations. A constraining factor with satellite technology, however, is the cost involved in using this technology. However, given its wide covering footprint it is a very good technology to employ in mass markets.

Wireless Local Access Networks

Wireless Local Area Networks (WLANs) have a large role to play in taking communications services to the mass market. Today, WLANs are beginning to evolve beyond their novel role of provisioning high-speed Internet access services to enterprise. WLANs are providing Internet access to business end users and high-end residential travellers that require service in localised hotspots such as airports, coffee shops, conference centres, hotels and so on.

WLAN can easily be deployed as a low cost RAN solution in sparsely populated rural areas simply by changing its radio capabilities, for example by using external antennae. WLAN is appealing to end users that are always moving around and want to use a laptop PC and/or personal digital assistant (PDA) that has been WLAN enabled make Internet connection and access the enterprise Intranet site. WLAN presents an interesting case because of its low bandwidth to cost ratio and because of its nomadic properties. WLAN is an excellent supplement GSM, CDMA, fixed line DSL and other broadband access technologies. WiFi is a very good example of Wireless broadband access. WiFi growth has been very good for a number of reasons such as its ability to satisfy the connectivity needs of end users and it ability to be profitable to operators and service providers.

BRINGING IT ALL TOGETHER

The main communications needs of end users in developing markets remains voice and basic data services (which is mainly SMS). Operators and end users alike view developing areas as offering very limited opportunities, for advanced high speed data and non-voice technologies. As we have argued before (see chapter 1), the provision of Internet access and other high yield value added services and applications (VAS) will create more opportunities for the poor by giving them access to information and knowledge which in turn will assist them to bridge the digital divide and become part of mainstream global economic development. Each developing mass market has its own profile, since the degree to which the different segments in the mass market differ depends on their level of ICT knowledge and technology needs with regard to mobility and a host of other factors.

Clearly, when it comes to identifying appropriate technology solutions for the mass market and rural poor, there is no perfect single solution for all developing markets and rural communities. However, we must point out that although there is no single solution for developing countries, there are a number of trends that these markets have in common when it comes to communications infrastructure roll-out. It is commonly known that wireless technologies are more investment friendly than wireline solutions, mainly because wireless technologies are easier and much faster to deploy that fixed line. This is largely because of the labour intensity involved when erecting a fixed line network. At the same time, wireless technologies enable a much wider area to be covered and require much less labour intensive civil construction work than fixed line.

The other generally held belief of communications operators and service providers is that the needs of end users in mass markets are for voice services. Hence, they view voice services as the area that will offer them the most growth using mobile communications. It is true that there has been an upsurge in mobile voice and basic data (SMS) penetration in developing countries during the last 2 to 3 years, but this has been mainly in the urban areas. Nonetheless, it is important to note that major opportunities are still available in other underserved developing areas such as the rural space. Serving these areas will bring more profits for telecom operators as their margins and customer base in urban markets come under pressure and begin to decline.

Each technology solution presents its own advantages and/or disadvantages. The main factor for consideration however, remains the radio/access network technology (RAN) that is deployed. A number of RAN technologies have surfaced during the past decade, each of these offering interesting deployment potential from both a business and technical viewpoint. Baldwin and Thomas (2005)[21] identify two major enablers of mobility at the infrastructure level, presently in place: (1) indoor wireless local area network (WLAN) and (2) outdoor (cellular). These technologies are suitable for data. GSM and CDMA technologies offer mainly mobile voice and data services through 2G and 2.5G. However, today more bandwidth

[21] See Baldwin and Thomas. 2005 (footnote 20).

and mobility is being offered via 3G, HSDPA, CDMA 2000 EVDO and other forthcoming technologies such as WiMax and 4G. These technologies offer a full suite of top communication services that include voice, data, broadband data, video and multimedia user applications.

WiMax technology is readily being seen as a saviour technology in much of the developing world. In fact, lower cost BWA promises to bring growth to thriving developing market economies like those of India and China. BWA is having a direct impact on these economies mainly due to its pragmatic short-term applications and longer-term influence on the development of the poor through the wider distribution of information and knowledge sharing. In many developing countries, broadband deployment is targeted towards urban and suburban areas and it is very possible that this trend will continue in the future. Poor availability of fixed line access remains a major barrier to the widespread take up of DSL. This creates a high need for alternative broadband technologies such a wireless broadband. WiMax technology is very well positioned to supplement this need. In fact, many factors such as the availability of fixed line, switch distance, backhauling costs and teledensity determine the choice of wireless solutions. A major advantage of WiMax is its wireless broadband nature and its ability to provision for access to developing regions where providing basic voice or broadband access using fixed-line service is not economically feasible option.

WiMAX standard, 802.16-2004, is a fixed variant of WiMax and can be deployed where operators have a need to provide backhaul in the cellular network. Furthermore, this technology can be employed to boost public access WiFi hot spots by means of enhancing throughput in the backhaul network. This makes it much more economical to deploy WiFi access points. Operators can reduce establishment capex by removing the need to use either copper or fiber for backhaul. At the same time, the multiple risks that can bring a halt to service such as vandalism or and cable theft can be avoided.

The technologies to provide widespread access to the poor in developing markets exist with the added capability to connect mass-market end users with the same user experience they would experience by being connected to urban broadband sites. Today, it is quite possible for operators to put together a mix of technologies using turnkey hybrid solutions that include satellite, WCDMA, Wireless DSL, WiMAX, GSM or WLAN to satisfy the communications needs of end users in mass markets that are economically and technologically viable.

Taking Mobile Voice into Under-Serviced Areas

Citing pyramid research, Baldwin and Thomas (2005) suggest voice-dominated trends will continue in the near future. Although there is an expected marginal decline in adding 2G subscribers to the network, from 93% during 2004–2005 to a projected 86% during 2008, 2G-service demand remains high.[22] Such trends mean that operators will direct their investments towards technology solutions that have a low CAPEX requirement and which will enable them to increase their margins by relying on voice and other high-yield applications such as international calling, mobile roaming and SMS. Beyond any doubt, cellular networks are the main facilitators for wide coverage mobile voice and data applications. Presently, 2G networks (GSM, CDMAOne, TDMA, and PDC/PHS) account for more than 97% of the total mobile market.

Technology Solutions for Rural Poor in Mass Markets

The mass market is an area that has the least mobile coverage. Mass markets include areas where there is little or no communications coverage such as in the countryside, villages, slums and other places where there is little population density and where the poorest of the poor mostly reside. Operators face the increasing challenge of bringing coverage to these areas at affordable cost. As we have noted so far, the two most suitable technology choices, for taking basic mobile communications services to the mass market are 2GSM and CDMA IS95.

These technologies can allow operators to offer voice and limited Internet access at low data bit rates. As is commonly known, the majority of mobile telecom operators have adopted GSM globally. It is widely recognised as the reference technology for 2G mobile telephony. Furthermore, GSM has a huge global installed user base in both the developed and developing countries. This gives GSM scale advantages to reduce costs. At the same time GSM offers end users a wide array of choice for handsets, as well as enabling mobile operators to take advantage of its low entry costs as a result of its high level of technical maturity. A major disadvantage with GSM is that it can be an expensive technology choice for use in very low-density and remote areas.

CDMA is also a cost efficient solution, competing with GSM. Operating at the 450 MHz frequency, it is recognised as an economically viable solution in outdoor conditions and low-density rural areas where there is less than 10% teledensity, mainly because of its superior geographical reach. CDMA's low frequency range provides the most relevant technology for use in remote and rural areas. This is because lower frequency bands provide more range and this means that economies of scale can be realised because fewer site installations are required, resulting in a reduction in

---

[22] See Baldwin and Thomas. 2005 (footnote 20).

CAPEX when compared to other technologies using higher frequency ranges.[23]

The economic advantages that can be gained from CDMA create opportunites for installing this technology in areas where there is low-density and income or gross domestic product per-capita. CDMA fully supports voice service, while offering an acceptable data rate (comparable to the lowest class of fixed DSL). One disadvantage is the problem of quality reduction when scaling up to serve an increasing subscriber base, especially in high-density areas. There is a limited choice of handsets (few providers of handsets with low end features) compared to GSM, and limited roaming ability for public mobile service applications is another constraint of CDMA.

Bringing in More Than Just Voice to Reduce the Digital Divide

According to the International Telecommunication Union (ITU), the last two or three years has witnessed the gradual narrowing of the digital divide between the developed and developing world mainly because of the growth in mobile and Internet penetration. However, there is still a real intra-country digital divide, in terms of the national availability of ICT services in urban areas that have been addressed and unserved rural areas. Reducing the digital divide is a challenge for both public and private stakeholders, if we consider that ICT is a tool for economic and social development. The diffusion of Internet-enabled value-added services and applications influences daily life and offers many new economic opportunities to individuals. Furthermore, considering the financial constraints of mass market end users, many relevant applications using wireless broadband Internet access technologies, such as satellite, 2GSM, 3GSM, WCDMA, WLAN and WiMax, can be used to connect poor communities.

Device Management – Using Devices to Spread Services

Now that the range of technologies to enable mass market communications exists, we need to establish what the main requirements are to enable connection of the poor in mass markets to global information networks. The first step to network connectivity requires access to basic tools such

---

[23] See Baldwin and Thomas. 2005 (footnote 20).

as a SIM card and handset or other devices. A SIM card is an intelligent microchip set that plugs into a device such as a mobile handset and allows the user to communicate with the network elements such as the BTS, MSC (main switching centre) and so forth using radio signals. For as long as the mass market does not have access to these basics they will never be able to participate in the global information society. However, the prices of these basic tools are prohibitive. Therefore, prices for these elements need to be reduced to open the information gateway for the poor in mass markets.

There have been a number of developments to reduce the prices for these tools. Increasingly, operators and service providers are realising that extracting value from the bottom of the pyramid needs a new focus on reducing or even eliminating access costs. This has led to a number of access price reduction developments, such as the promotion of refurbished handsets and providing SIM card packages bundled with airtime packages for little or no charge. For example, in South Africa, the mobile operators MTN, Vodacom, Cell C and Virgin provide SIM card and airtime bundled packages for less than one dollar. There has also been a lot of development toward reducing handset prices for both prepaid and postpaid contracts. At the same time, a number of mobile phone manufacturers such as Motorolla, Nokia and Sony Ericsson are teaming up to produce handsets that costs less than one dollar, making them affordable to the mass market in developing countries.

These days, any PC can be communications enabled using WLAN, GPRS, 3G and HSDPA connectivity simply by adding a network communications card to the device's hardware. 3G network cards cost less than US$50 and are readily available from operators and service providers. At the end of 2005, almost 90% of all PCs sold in the world were WiFi-enabled with WiFi communication chipsets. All personal digital assistants (PDAs) available today come standard with built-in GPRS and WLAN connectivity.

Until now, potential end users in developing markets have missed out on the innovative developments taking place in mobile services. In the following sections we present and discuss some of the main mobile service innovations that would benefit the mass market end users and increase the quality of their lives.

INNOVATIONS IN MOBILE COMMUNICATION SERVICES

Mobile communications service innovations have been gigantic. Service developments have taken place in the following major areas:

- Voice
- Messaging
- Infotainment
- Internet access
- Location based services
- M-Commerce
- Mobile media participation
- Video services

We touch on these services and attempt to show how the mass market can use some of these services.

Voice

Voice is the most basic of communication services and is one of the best known services that has the widest application for mass market end users. By being able to use voice to communicate, mass market end users will be able to keep in contact with friends, family and business contacts enabling them to substantially improve their social and business relationships. Voice communication facilitates economic growth by empowering end users to cross over the barriers that prevent them from accessing basic information.

Incoming Call - Services

Almost all operators offer services that assist their customers to manage their incoming calls. An example of such a service is *call divert* which allows the end user to divert incoming calls to another phone or location. This service is very beneficial to mass market end users because it prevents the loss of important calls received by the end user. For example, a person operating a small or microbusiness may lose an important business opportunity because they fail to act on a call immediately. By diverting the call to another number, someone else can take the call and instantly act on the business opportunity.

Voicemail

This is an important service that works like a traditional answering machine. If for any reason, the users phone line is engaged, or the incoming call is not answered within a prescribed time, the called party's mobile operator diverts the caller to a voicemail system. Here, the calling party is prompted to leave a message. This function is enhanced by a text message or in some cases a call from the operator informing the end user that they have a message waiting. The end user can then retrieve the new message by dialling a number provided by their operator. Voicemail services allow the user to listen to saved messages, save the message for a restricted time period or simply delete it. Such applications are very important for mass market end users where every incoming call is important, especially business calls. Without this service, entrepreneurs and businesses operating in developing countries lose out on lucrative commercial opportunities.

Call Waiting & Call Hold

The mobile handset can be set to alert the end user on all/any incoming calls. This allows the user place an existing call on hold by pressing a switch on their mobile phone and to take the incoming call. By simply flicking a switch, the user is able to retrieve the call on hold.

Call Forwarding

Incoming mobile calls can be diverted to another phone number.

Call Restrictions

Call restrictions allow the end user to restrict certain calls being made. For example, if another person uses the phone, it can be programmed to restrict outgoing calls. Most operators allow users to set up this function on their phones to prevent other users using their phone, to make calls to international or premium-rated numbers.

Caller Line Identification

The majority of mobile phone operators enable the number of the party calling to be displayed on the receiving phones screen. However, this function can be disabled using the handset. This service is important for mass market end users because it allows them identify and differentiate between important and less important incoming calls.

Push-to-Talk over Cellular

A relatively new operator offering is Push-to-Talk over Cellular (PoC) services. This innovative service allows end users to use their phones like walkie-talkies, by simply pushing a button on their phones to talk to another user or group of users. This service allows only one person to talk at a time and all other participants can hear the conversation.

Messaging

Messaging is another innovative group of service developments. Using the messaging function on the mobile phone, users can send messages across the mobile network to other users. Messaging can be a very valuable tool for end users in mass markets.

Short Messaging Service

Short Messaging Service (SMS) enables end users to send and receive text messages on a mobile phone. End users use the numbered keypad on the mobile handset to input alphanumeric characters. A single SMS can consist of a message with up to 160 characters and can be sent to users of different operator networks. All the available mobile phones on the market today support SMS. SMS has taken off in global markets and it is estimated that billions of text messages are sent across the globe each week. SMS has tremendous benefits for end users in mass markets. For example, taxi drivers in Nigeria rely on SMS to receive instructions from their control rooms of where to pick up their customers. Another good example is the use of SMS by the thousands of people living in informal settlements in South Africa, Nigeria, Indonesia and Brazil to keep in contact with employers, business associates, family and friends.

Besides person-to-person SMS there are also a wide variety of text messages that are content-based. Most operators offer subscription that allows subscribers to receive news, financial market information, sport and entertainment content directly on their mobile phone in SMS form.

Multimedia Messaging Service

Multimedia Messaging Service (MMS) is a store and forward messaging service that allows mobile subscribers to exchange multimedia messages with other mobile subscribers. As such, it can be viewed as an evolution of SMS, with MMS providing support to the transmission of more advanced types of media such as text, picture, audio, video or multimedia. The end user can simply create a multimedia message by using the built-in phone camera or an accessory camera. They could also use retrieve other images and sounds that were previously stored in the phone in the phone's memory. Like SMS, if the receiving party's phone is switched off, the multimedia message will be stored on the network and sent as soon as they switch on their phone.

Depending on the type of handset used, a number of multimedia messages can be stored in the users handset and reviewed or forwarded at any time. MMS has the added advantage of being able to send and receive multimedia messages from one phone to another as well as from the phone directly to email. However, to be able to send or receive MMS messages, the users phone must be compatible and able to use a GPRS or 3GSM network. MMS is a very valuable service for end users in mass markets, but the cost to access and use this service is high. Price reductions for MMS will most certainly result in widespread adoption of this service among mass market end users.

Instant Messaging

Instant Messaging (IM) is a real time application that facilitates the exchange of written communication between people using a PC or laptop using Internet Protocol (IP). Mobile IM enables users to participate in instant messaging services using their mobile phones. Using mobile IM, users are able to address messages to other users using an alias and address book. Mobile IM allows the sender to know when their friends are available on the network. Mobile IM is advantageous in that messages are sent and received in real time through mobile handsets in much the same way as fixed IM services. To use these services mobile users need to have an active IM account like those offered by AOL and Yahoo. They also require a compatible handset that has messaging client software preinstalled and that is capable of running over a GPRS or 3GSM network. Mobile IM presents an inexpensive solution for mass market end users that have a need always-on real time written communication.

E-mail

Wireless e-mail allows users to send and receive e-mail across wireless networks and devices. Network development involving 2.5G (GPRS) and 3GSM providing always-on connectivity enables users to access to their e-mail at any time. Nowadays, there is a large selection of mobile handsets available, from mid-priced mobile phones through to smartphones and specific e-mail devices such as the Blackberry, that can support wireless 'push' e-mail services. Push e-mail technology enables e-mails to be sent directly to the mobile handset as soon as the e-mail server receives them, instead of waiting for the end user or e-mail client to request the email. It is important to note that wireless e-mail services require devices loaded with software that is capable of supporting this application.

Infotainment

Infotainment is an area of mobile service development that holds great promise for mass market adoption in developing countries. There have been a number of infotainment developments such as:

- TV/Video
- Music
- Gaming

TV/Video

Ericsson is one of the industry players that believe mobile TV and video services will become a very profitable, highly proliferated market within the next few years. Theoretically, mobile TV services provide exactly the same 'live' broadcast images like those that are viewed over the television. The convergence of TV with mobile phone systems can be done in two ways (1) real time streaming over 3G networks and (2) making use of dedicated mobile broadcast networks.

Broadcast services streamed over 3G networks make use of the increased capacity offered by 3GSM technology. These can only be viewed using a 3GSM mobile phone. In developed countries, a number of global operators such as Hutchison and Vodafone are forging ahead with services that stream TV channels to 3G handsets.

Another way that operators can provide TV services is by using specially dedicated mobile TV broadcast networks that make use of specialised technologies such as digital video broadcast-handheld (DVB-H), digital mobile broadband (DMB) and media forward link only (Media-FLO). Most broadcast mobile TV trials in Europe have been based on DVB-H. Operators such as KPN Mobile, Telefónica Móviles, MTN, O2 and Telstra have been involved in DVB-H trial technology.

SK Telecom's TU Media subsidiary and Japan's Mobile Broadcasting Corporation have been experimenting in South Korea with another mobile TV broadcast technology called satellite digital multimedia broadcasting (S-DMB). This technology uses a single satellite. Launched at the end of 2005, this service already has an uptake of 1 million users. European operators, BT and Virgin Mobile have joined forces to establish BT Livetime which uses DMB and is in turn based on the DAB technology used for digital radio. In Germany, trials are being conducted for mobile broadcasting using terrestrial digital multimedia broadcasting (T-DMB).

Multimedia broadcast and multicast services (MBMS) are technologies that have been engineered to efficiently deliver multimedia content over 3G networks. Operators using MBMS, could achieve the same level of coverage without the need for incurring expenses to buy additional spectrum, licensing or base-station construction.

Today, mobile TV is commercially available and industry expectations are that by 2010, some 60 million plus users will be using mobile TV services, worldwide. The hunger for information and knowledge of mass market consumers makes mobile TV a very lucrative service, but once again the price charged for this service needs to be brought down to affordable levels, to draw in mass market end users in developing countries.

Music

Both, GSM and CDMA technologies allow operators to provide a wide selection of music services to end users. There is a huge demand for portable music in both the developed and developing markets. At the same time new developments in the area of content development within the mobile industry has sparked off a growing list of mobile music products for end users. Until recently, mobile music supply was limited to the provision of ringtones.

Ringtones are small audio files that are played by the mobile phone to alert the user to an incoming call. Ringtones may be made up of a few notes of familiar musical tunes. The traditional purchasing process for ringtones involved the user making a call using an Interactive voice recognition (IVR) system. These days most ringtones are directly accessed and downloaded from wireless application protocol (WAP) portals and are then sent directly to the requesting user's phone using premium rate SMS. In other words, the end user pays for the ringtone through the price billed for the SMS.

'The mobile music market in China already generates more revenue than the region's traditional music format market.' (Wang Jianzhou, President – China Mobile)[24]

The latest developments in ringtone products are ringback and truetones. Ringback tones are sounds that calling parties are able listen to over the line when they call another party. Instead of hearing a ringing tone at the other end they can hear their selected ringback tone. Realtones, also known as truetones or mastertones make use of real sound recordings. Realtones can categorised into two main categories (1) tones that make use of a recording that has been commercially released by a record company and (2) tones that make their own recordings.

Besides the developments taking place in toning, other developments that enable users to access full music tracks and store them onto the memory of their mobile device is another innovative development. Mobile phones are available on the market with built in synch-software and MP3 technology, as well as other digital music format playback capabilities. Realtones and full track downloads are available from Internet and operator portals.

[24]Wang Jianzhou (President, China Mobil). 2006. Speech made at 3GSM World Congress, Barcelona, February 2006.

Gaming

There have been some strong disruptions in the mobile gaming market that has quickly evolved in recent years. Early examples of mobile games such as Snake and Chess were built directly into handsets and displayed over a monochrome mobile screen. The new set of handsets available today provides pre-stored games that rely on the phone's colour screen to display them and inbuilt sound and memory features for high quality sound and storage. The vast array of newly developed inbuilt games that are available today are dependent on the make and model of the handset. Table 5.3 shows the Ericsson forecasts for mobile gaming.

Mobile gaming is expected to reach about 347 million downloads by 2011. New wireless network technology developments such as GPRS and 3GSM services have promoted the development and delivery of much more advanced mobile games. Users can download these by paying a fee or directly using the operator's portal or the Internet.

Mobile gaming involving multiple players is another relatively new development. Multiplayer gaming is a growing market. A number of content developers through content providers offer games that allow as many as eight people to play in the same game in real time across the mobile communications network. 3GSM services have seen the introduction of specialised tariff packages such as unlimited usage. Such innovations promise solid growth for multiplayer gaming which, is expected to grow much faster than single player gaming.

Mobile Internet Services

The majority of Internet web pages are not compatible for viewing on the small screen of a mobile phone. As result, mobile operators make use of WAP technology to provide users with access to Internet services and

Table 5.3 Mobile gaming market growth forecast 2005–2011

Year	Number of downloads (in million)
2005	76
2007	163
2009	265
2010	300
2011	347

Source: Ericsson, 2006[25] with permission

[25] Ericsson global estimates. 2007.

sites. Users of this service need a WAP enabled phone to view the hundreds of thousands of WAP sites that contain an abundance of information and images that are to be found on the Internet.

Both software and hardware technology developments for handsets have witnessed a variety of new phones on the market that have large colour screens and full Internet browsing capabilities. This enables them to interface with standard web pages. The GSM association has partnered with a number of industry players such as Nokia, Vodafone, Microsoft and others to develop mobile phone Internet access sites. These web sites addresses end with the suffix .mobi, instead of .com or .org and allow users to differentiate that the site can be view on a mobile phone.[26] The Internet can also be accessed through the mobile network, using a laptop computer or personal digital assistant (PDA). Such developments are great to facilitate the quick proliferation of Internet access to the poor in mass markets.

Location-Based Services

Location-based services (LBS) are personalised mobile services that users can use to access specific information, for example: information on hotels, restaurants, shops, maps and so on. Advances in wireless mobile technology provide the possibility to distinguish between different methods of location tracking. For example, using Cell-ID, a number of LBS applications are able to determine the users location. With Cell-ID, the cellular network intelligently identifies the closest cell through the home logic register (HLR) or visitor logic register (VLR) in the BTS the user is connected to. This simple form of location tracking is fully supported by all GSM and CDMA phones.

The accurate tracking of the user's location depends on a number of factors such as line of sight obstruction, but generally mobile devices can be tracked to an area of between 200 metres and one kilometre. Another highly advanced and precise location technology is global position system (GPS). Relying on satellites, GPS is able to pinpoint the exact location of a mobile device that has been equipped with special hardware and software to receive signals from the satellite. This process is called triangulation and is illustrated in figure 5.10. By receiving signals from three of the satellites, a device equipped with a suitable chip can pinpoint its current location anywhere on earth to within a few meters.

Assisted GPS (A-GPS) systems shown in figure 5.11 have therefore been set up to resolve the long delay that can occur in locating a unit.

Data about the mobile unit is transmitted through the network of base stations to speed up the process of location, bringing it down to only a

---

[26] See www.dotmobi.org/ last accessed on July 3, 2007.

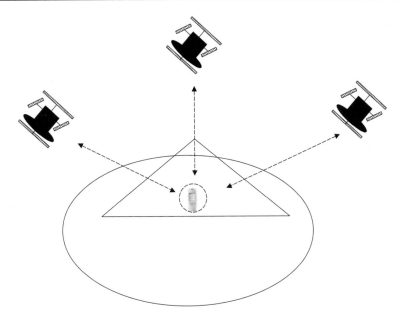

Figure 5.10 Illustration of LBS triangulation

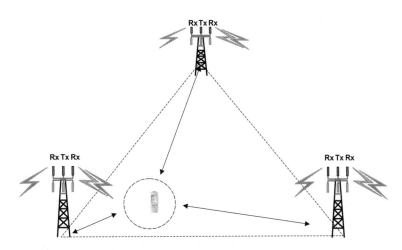

Figure 5.11 Assisted GPS location based service

few seconds. This advanced technique is being introduced in new GSM handsets. A-GPS is a well-proven technique already used worldwide for the accurate tracking of valuable assets such as shipping containers and high-value loads.

204 MOBILISING WIRELESS COMMUNICATIONS FOR MASS MARKETS

User-Generated Content

Content that generated by end users themselves is called user-generated content (UGC). UGC is available in a variety of media formats such as data, music, video and multimedia. High speed networks such as 3GSM and CDMA 2000 have facilitated the delivery of user-generated multimedia content.

UGC includes common content such as jokes and witty questions that users generate and then submit to the network via a host for the use of other users. Using the mobile to is another popular way for users to exchange views and share information.

Using UGC end users could develop personalised music which may be retrieved by other users. A good example of this for of UGC is the case of China Mobile. During October 2005, China Mobile introduced an innovative competition inviting mobile users to generate their own music and submit it for the competition. This turned out to be so successful that more than 500 000 mobile users entered the contest. This music is now stored China Mobile's portal. Mobile users can vote for their popular music artist using SMS, WAP, IVR and the Internet. Furthermore, they can also download the songs directly from China Mobile's portal and store them on their mobile phones or, if they prefer, use these tones as ringback tones.

UGC is beneficial to both the user as well as the network operator. The advantage to users is that they do not have to pay any copyright fees to distribute or download this content and the content is much less expensive for them to buy and distribute than songs from established warehouses. Another new development in UGC is user-generated multimedia content services such as 3UK's, SeeMeTV. This service was launched in October 2005 and provides mini-movies that have been video recorded by 3's customers using their mobile phones. These amateur videos are then sent to mobile operator using MMS where they are uploaded and stored until it is retrieved and shown on TV.

Citizen Journalism

Citizen journalism is a relatively new mobile development that has been brought about by an increase in the use of mobile handsets to generate, share and distribute photos and videos. This has enabled mobile users to participate much actively and directly in the world of journalism where they can report and distribute the latest news and information. In most instances, mobile users arrive by chance on a scene where they are able to capture events on their mobile phones. They then record the events and can then send these directly to a news agency or portal where others can view it.

Many news broadcasters such as the BBC, Al Jazeera, NDTV and CNN will welcome such content. These amateur videos and pictures are generally of a high quality and suitable to be broadcast over television.

M-Commerce

Mobile commerce involves the use of a mobile device to transact electronically. The mobile phone provides a personal, convenient, immediate, seamless and secure way for end users to transact. M-commerce also enables businesses to broaden their market, enhance service offerings and streamline their costs. Operators also benefit from m-commerce because there is a high need for them to defend and grow existing average revenue per user (ARPU). M-commerce promises to be a very lucrative offering that could assist them in achieving this objective. Some m-commerce service examples include paying for purchases at cash outlets of large supermarket chains and paying for movie tickets. These types of payments are known as micropayments and involve transactions that have a total single transaction value of less than US$10. Expectations are that future growth will come mostly from demand for digital content and services that are valued at less than US$10.

Other m-commerce services include financial services such as m-banking. M-banking allows mobile users to check their bank accounts, pay accounts, buy airtime and so on. M-banking services have proven to be very popular in developing countries such as the Phillippines where the banking sector has been very poorly rated. The use of mobile-enabled commerce services could easily address major gaps in developing countries such as the Philippines that are crucial to support their social and economic development. Most developing countries, especially rural areas have very limited access to financial services. As a result, a big percentage of the rural population use cash to transact. The advances in mobile technologies and services in developing countries can be a unique platform to offer financial services. This platform provides excellent opportunities to foster closer synergies between operators and financial service providers such as, for example, Grameen Bank and Grameen Phone in Bangladesh.

Case Study 2: *Micropayment systems and their application to mobile networks – An assessment of mobile-enabled financial services in the Philippines*

The proliferation of mobile communications in developing countries has the potential to bring a wide range of financial services to an entirely new customer base, according to a new report commissioned by infoDev in partnership with the International Finance Corporation (IFC) and the GSM Association. The report, which focuses on the use of mobiles for micro-payments in the Philippines, found that mobile-enabled financial services, or m-banking, can address a major service gap in developing countries that is critical to their social and economic development.

In the Philippines, 3.5 million people are using a service that allows them to transfer money over the two major mobile networks operated by SMART Communications and Globe Telecom. The experience in the Philippines shows that mobile-enabled financial services have the capability to bring advantages to all stakeholders. For users, there is an opportunity to become engaged in the formal banking sector, to facilitate and reduce the costs of remittances and to enable financial transactions without the costs (travel) and risks (theft) associated with the use of cash. Here, we list some other advantages:

- For operators: a significant increase in text messaging revenues and a large drop in customer churn.
- For consumers: m-commerce is more secure and flexible than cash, allowing consumers to make payments remotely.
- For banks: an increase in their customer reach and the added cash float available to the bank.
- For retailers: added business opportunities through the sale of prepaid account credits.
- For microfinance institutions: the ability to advance funds into remote areas and have regular repayments that do not significantly inconvenience the user.
- For service industries and utilities: the ability to get payments electronically from a significant portion of the overall population.

Source: Information for Development Program[27]

Video Services

Today's 3GSM networks enable the streaming of highly enriched multi-media services such as video sharing and video calling that make use of live camera or recorded video clips to optimise the end user's mobile communication experiences and at the same time make possible a real time method of introducing pictorial elements into a live phone conversation.

---

[27] Information for Development Program. 2007. http://www.infodev.org/en/Publication.43.html last accessed on July 14, 2007.

Video Sharing

A video sharing service allows mobile phone users to share a live camera view or a video clip while they are speaking on their video-enabled mobile phone. A video sharing service will usually commence as a two-way traditional voice call, with one user then able to add video interactivity to the call. Operators expect video sharing to be a highly popular service used by people to exchange everything from live video postcards on holiday to footage of a school sports day to video of a house for sale.

Video Calling

A video call originates as a connection between two video-enabled mobile phones where both people on the call see as well as hear each other in real-time. It is a live video feed caller-to-caller.

OTHER SERVICE INNOVATIONS

Besides the services mentioned, there are a number of other innovative service developments that require mention. These are briefly discussed next.

Airtime Transfer

Airtime transfer refers to a mobile service that allows users to transfer airtime credits to other mobile phone users. This service has great potential for mass market users. Already, in some developing countries of Africa such as South Africa, Tanzania and Kenya, mobile users are using airtime as a form of currency to transact. This is only possible because they can transfer airtime to other users, in return for cash.

Increasing Customer Loyalty

Customer loyalty is a function of perceived value. Operators and service providers that want to increase customer loyalty for both post-paid and prepaid customers can do this in the following ways:

- Package free airtime bundles with prepaid usage, for example where a prepaid/post-paid customer has been loyal to an operator the operator can give this customer free airtime minutes
- Provide free gifts to customers according to the number of airtime units used, for example where customers use 500 minutes they could

receive a small appliance. If they use 1000 minutes they receive a larger appliance or shopping voucher.

CLOSING ARGUMENTS

For operators, the adoption of GPRS, CDMA 2000 1X, WiMax, WLAN or satellite is a fast and cost effective strategy that not only supports the real first wave of mobile Internet services, but also represents a big step towards 3G or fourth generation (4G) networks and services that will be adopted in future by the developing mass markets.

The most noteworthy aspect of these development pathways is that each has enabled operators and service providers to make the highest order of information products and services available to end users, but a lot still needs to be done to take these products and services to the poor in mass markets. Also, it is important to realise that, in keeping with the traditional view on value creation, these products and services have been reserved for the more affluent customers in both developed and developing countries. But all is not lost. The technology is available to provide ubiquitous communications solutions for all. The only barrier that lies between the 'haves' and the 'have-nots', is for communications operators and service providers to find a way to reach the masses without compromising their business objectives. New business models are urgently needed for this.

Traditional business logic about value creation is to be found in the thinking that the more rare a commodity, the higher is its value. In the networked world, this logic has been reversed. The more nodes that get connected to the operators network the more valuable the network becomes. This means that there is a powerful business case for operators to partner with the right stakeholders and to undertake communications investment in developing country mass markets. In this way, mass markets will be able to take up the different mobile service innovations that have been created.

CONCLUSION

There is an increasing realisation that providing communications to mass markets is the key to sustainable future global economic growth. Today, over two billion people that share our world live without the simple basics that are so important to support a comfortable and healthy lifestyle. In this chapter, we looked what still needs to be done to ensure mass market inclusion in the global marketplace. We attempted to show who the main drivers are or should be and also looked at the some of the main role players that should promote communications in mass markets. We then went on to discuss some of the main wireless technology developments such as global services mobile (GSM) and code division multiple access (CDMA) and WiMax to show what wireless technologies are available today to operators to roll out communications infrastructure in mass markets.

Some further communications equipment developments were also highlighted, such as micro base stations, co-siting and refurbished communication equipment that we think operators could use to assist in spreading communications to mass markets much more cost effectively. The broad implications for increasing teledensity diffusion, using low-cost handsets and SIM cards, were also briefly discussed as a means to extend communications to mass market end users. Finally, we concluded this chapter by discussing a number of mobile service developments, such as mobile TV, messaging, infotainment, mobile Internet, video services and other innovations that included airtime transfer, increasing prepaid customer loyalty and simplified pricing and we attempted to show how these could be applied to the mass market. In chapter 6, we look at business models for Internet based services in rural areas and the role of development institutions.

6

Defining Innovative Business Models for Sustainable Telecoms Growth

OVERVIEW

The penetration levels both fixed and/or mobile phones in the developing countries, particularly in rural areas, are still very low and, on average, consumers today are paying a lot per minute of usage for the quality they are receiving. Businesses are suffering for not having a proper telecommunications infrastructure. The rural telecommunications market is evolving faster than ever and the majority of next three billion users live in rural areas of developing economies. Entering this market with traditional business models will prove disastrous for the profitability and sustainability of the telecom services providers. Cutting corners and offering lower grade services will not excite the users and will have a negative impact on the development of the market. The increasing focus on profitability per unit and the ability of each service provider to generate acceptable free cash flows present challenges. The rural telecom services market is so far driven by the success of prepaid tariffs. This segment will quickly become unprofitable if the costs are not controlled adequately. Indeed, in some cases, prepaid already shows negative net present values. The average subscriber ARPU level is declining and will continue to decline over next few years as the subscriber base expands at the bottom of the consumer economic pyramid. This will have a negative impact on average subscriber valuation. However, the subscriber growth patterns

Business Models for Sustainable Telecoms Growth in Developing Economies
S. Kaul, F. Ali, S. Janakiram and B. Wattenström
© 2008 S. Kaul, F. Ali, S. Janakiram and B. Wattenström

and forecasts are amazing and the operators are anticipating huge growth and therefore, a great need for network capacity, while sustaining and optimising the operational frameworks enabling lower OPEX and better cost of management.

Business model innovation is crucial. Operators need to make some fundamental changes to their organisations and business designs as part of their innovation initiatives and an examination of their financial performance suggests the reason for this. When we looked at financial performance over a five-year period, we found striking differences across different types of innovation. Business model innovation had a much stronger correlation with operating margin growth than other types of innovation.

Looking across the top actions business model innovators were taking, we found that companies innovating through strategic partnerships enjoyed the highest operating margin growth. As one CEO from Botswana remarked, 'reducing the cost base through cooperative models is important for any growth strategy.'

PARADIGM SHIFT IN STRATEGIC THINKING

If we analyse the last five years there has been lot of focus on bringing in high ARPU subscribers and that has definitely hampered penetration levels. But, despite the fact that operators have been strategically focusing on high value subscribers, particularly mobile subscribers, turnover has exceeded all forecasts.

However we now see a paradigm shift in the strategic thinking which is the increased focus towards mass markets with the intention to penetrate in so call low-end market segment, to bring in as many users as operators' network capacities and resources will allow. If this thinking continues to prevail operators in the region will need to adopt business models that will enable operators to cope with the growing network capacity demand and help build network infrastructure and operational frameworks which are optimised and efficient to sustain and catalyse the envisaged growth.

SUSTAINABLE SUBSCRIBER GROWTH

The shift in strategic thinking will have a dramatic impact on the subscriber growth outlook and the active subscriber counts. Operators in developing economies, particularly mobile operators, have been growing much faster than their peers in the developed economies, with the slowest growing players recording a compound annual growth rate of up to 38% between the years 2001–2007.

The rise of regulatory bodies and expansion of mobile and fixed telephony license awards make developing economies more stable for further growth. This strong subscriber growth is sustainable for at least 5 to 7 years.

This continued growth in subscriber numbers offers a significant opportunity for all the important stakeholders in the African telecommunications industry, governments, telecom regulators, network operators, service providers and telecom suppliers. The key task is to analyse the business models that can address the challenges that this sustainable growth is expected to bring.

GIVING ACCESS TO RURAL AREAS: KEY TECHNICAL CONSTRAINTS

Developing access in rural areas is a critical and strategic challenge for policy makers of developing countries, since telecom operators have historically spent their efforts on urban areas. Numerous technical options enable telecom suppliers to provide ICT services to rural areas, nevertheless access and core networks have to be considered together when focusing on rural connectivity. In effect, there are technological requirements to consider before focusing on access, which is just a part of the infrastructure roll-out. In fact, the quality and relevance of the technology chosen for the provision of access to rural areas will be influenced by other network considerations like existing backbones, quality of core network infrastructure, sufficient number of Internet Exchange Points (IXP) and backhauling options depending on existing telecom infrastructure and core network issues must be considered concurrently when focusing on rural connectivity.

The Access Issue

Fostering ICT development in rural areas of sub-Saharan Africa is a great challenge and the bundle of chosen solutions should be cost efficient to allow Total Cost of Ownership (TCO) reduction for telecom operators while also guaranteeing profits. There is no unique model designed to fit operators' constraints. Several factors influence the choice of access technologies for a targeted rural area, each region and country having different characteristics, physical constraints and existing infrastructure.

Core Network Issues

Today, most networks are designed to provide services relating to a specific application such as:

- Public Switched Telephone Networks (PSTN) that were originally designed to carry voice communication applications
- Data networks such as Internet Protocol (IP) networks that provide Internet services like access to the World Wide Web (WWW) and e-mail
- Mobile networks that provide mobile communication applications
- Cable networks that were initially developed to offer television distribution services and now deliver Internet access and voice services at reasonable cost.

Existing telecom operators hoping to provide access to rural areas will often build on their existing core network infrastructure. If they want to maintain quality of service and attract new customers, they have to upgrade their existing infrastructure (hardware and software) to efficiently manage and absorb an increasing customer base with its associated voice and data traffic. When investing in rural infrastructure projects, most service providers do not pay enough attention to the notion of quality of service continuation, as they try to optimise their revenues to absorb initial investments (CAPEX) and operational expenditures (OPEX), to achieve faster return on investment (ROI).

Transmission

In a telecommunications network, transmission is the transfer of information from one network point to another. The medium used may be copper cables, optical fibers or radio relays. Optical fiber and microwave may be found in urban areas, while microwave and satellite may be found in rural areas. Backhauling technology, which can be terrestrial (radio or cable) or non-terrestrial (satellite), enables voice and data services to be transmitted between core and access networks, while an access technology enables end users with a customer premises equipment (CPE) to connect to the network via an access point (BTS, WiFi hotspot, etc.), also known as last-mile technology.

Backbones ensure voice and data transmission between national and international networks. The backhauling issue is a major constraint for ICT infrastructure roll-out in rural areas, due to the long distance between the core network infrastructure, normally located in an urban area, and local access points.

The Importance of Backbones

A national backbone network is the infrastructure needed to allow broadband access to basic telecommunications and advanced value-added services for fixed and mobile telecom operators and ISPs. National backbones offer cost efficient access opportunities for intra-country communication and data exchange, rather than leasing foreign or external facilities. Long term return on investment on such projects requires some sort of public support, consisting of regulatory improvements and financing options. Public-private partnerships (PPP) involving governments and private players could then be considered to ensure financing of these national backbone networks.

National backbone development is crucial to ensure rural connectivity. There are still many SSA countries that lack a national backbone network, relying on other countries for connectivity. This results in high service delivery costs and limited opportunities for scalability.

How to Ensure Backhauling

There are different ways to ensure backhauling in rural and remote areas. Depending on the distance between core and access facilities, and the density of the target population, it is possible to choose between wired, radio and satellite backhauling solutions. Wired backhauling may be not appropriate for rural coverage due to heavy CAPEX costs. However, radio technologies and satellite enabled applications will have an impact on the costs and time required to extend the network, with a decrease in CAPEX thanks to reduced civil works and engineering, but increased OPEX especially for satellite. Costs can be particularly high in remote rural areas.

Complex Environment

Rural areas are often harder to serve due to harsh environments and physical remoteness. This makes civil works and engineering harder to manage, generating additional costs and longer time schedules for infrastructure rollout. The climate in SSA also affects equipment design; high temperatures can affect the efficiency of equipment. Theft and vandalism are threats when installing hardware on isolated sites; it is sometimes necessary to secure sites with dedicated security staff, adding further costs.

Lack of Infrastructure

Rural populations often lack basic infrastructure in terms of housing, public services (health, schools, post offices), copper equipment (fixed telephone lines), permanent power supply, roads, etc. These shortages make investment less attractive and complicate infrastructure roll-out.

Human Capability

Human capability is also a constraint, as most potential users do not know how to use new technologies, and need to be educated to do so. Illiteracy is high in developing countries, especially in rural areas of SSA. A local workforce able to install and maintain the network is also of great importance; however, this skilled labor is hard to find and to keep, due to high turnover.

High Entry Costs

The consequence of these constraints is high entry costs due to long distances, difficult access, and high transmission and civil works costs. Initial investment is high, but the future upgrade potential could be substantial, thanks to the possibility of remote software upgrades with the new generation networks.

In effect, network expansion/evolution sometimes does not require costly re-engineering. The cost of acquiring rural customers is also higher than for urban customers, due to the need for specific marketing campaigns and messages. However, franchising systems can decrease entry costs, for example by using a private local entity for customer care, or buying bandwidth at wholesale prices, as in the Grameen Phone business model.

Low Profitability

In terms of customer profiles, rural users often have, for the most part, lower incomes than their urban counterparts. The absence of human capability could also affect profits; it may take a long time to achieve full ICT awareness and empowerment. As a result, return on investment will take more time than in urban areas, as people will need to be trained and convinced about the clear benefits of using communication tools. Operators must count on large volumes of subscribers to compensate for the lower spending of these low-income users, although they can take into account incoming call opportunities, which represent 50–60% of rural telephony traffic.

Low Density Area

The business case is complicated by the fact that user density is insufficient to offset the high installation costs, resulting in low optimisation of base transmission sites (BTS) sites in the case of mobile telephony. A decrease in the number BTS sites needed to reach a certain number of users will affect Total Cost of Ownership and help to reduce costs per subscriber.

Understanding the Low ARPU Segment; End User Constraints

Service providers should be aware of the rural users' ICT needs, and adapt their distribution and marketing strategies accordingly. The profile of potential users and promoting the benefits of ICT, and training in their use has to be taken into account.

Case Study 1: *Grameen Village Phone Programme*

Grameen's Village Phone (VP) programme in Bangladesh is a well known case illustrating how telephone service can be extended to low-income rural dwellers. VPs are public access GSM phones that are owned and operated by local women entrepreneurs in villages throughout Bangladesh. Financial assistance for purchasing a GSM phone is provided by Grameen Bank, a microfinance institution. Once an entrepreneur has acquired a phone, she can then offer mobile payphone service at her shop, home or at the local market. Bulk airtime is purchased by the non-profit VP sponsor, Grameen Telecom (GTC) at a discount from the for-profit GSM operator Grameen Phone (GP). The airtime is then resold to the VP operators at a rate significantly below retail tariffs.

A US$30 Handset for Emerging Markets

Following a request to phone manufacturers from the GSM Association to develop and supply the next ultra-low cost handset to connect the unconnected in developing markets, Motorola was selected and said that it will release a US$30 mobile phone for commercial distribution in early 2006. This new generation of mobile phone for emerging markets will still allow the supplier to make a margin while being reliable and having an improved battery capacity since users may have limited access to electricity.

A US$100 Laptop for Children

MIT Media Lab is currently developing a US$100 laptop, a project that could revolutionise the education of the world's children, especially in emerging countries. The proposed machine will be a Linux-based, full-colour, full screen laptop, will use innovative power (including wind-up) and will be able to do almost everything except store huge amounts of data. The rugged laptop will be WiFi- and cell phone enabled, and have several USB ports. Its current specifications are: 500 MHz, 1 GB, 1 Mega-pixel. The idea is to distribute the machine through ministries of education willing to adopt a policy of 'one laptop per child.' Initial discussions have been held with China and Brazil. Initial orders will be limited to a minimum of one million units (with appropriate financing). The preliminary schedule was to have units ready for shipment by the end 2006 or early 2007. Although plagued by its own set of problems the program is gradually making inroads into developing countries.

Need for Low-Price Terminals

New generations of mobile phones developed for emerging markets are characterised by innovative design in terms of looks and functionality. The key drivers should be low price, basic features, and battery life. High illiteracy rates, especially in rural areas, should also be taken into account in the design of handsets, making appropriate use of voice recognition and icons. Prahalad and most telecom analysts suggest, when talking about bridging the digital divide, that the mobile phone has more potential than the personal computer.[1] In the near future, emerging markets will be wireless mobile-centric and not PC-centric. Still, the potential impact of broadband Internet access on local communities must be considered, especially for specific applications like e-education, e-health and e-governance, as in the five case studies described earlier. It would therefore make sense to give connectivity to public facilities like schools, health centers and post offices.

Provide Microfinancing Solutions

One of the main obstacles for mobile expansion among low-income users is the acquisition cost of the terminal. There are successful examples of

[1] Prahald, C.K. and Hammond, A. *What Works: Serving The Poor, Profitably, A Private Sector Strategy for Global Digital Opportunity.* Markle Foundation. WRI.

telecom operators such as the example of the Grameen Phone/Grameen Bank partnership that provided microfinancing solutions to allow end users or intermediaries to acquire a mobile phone and set up businesses.

Use Appropriate Distribution Channels: Reseller Network

Telecom operators should also develop a reseller network in rural areas to reach the widest number of potential users through phone shops, retailers, telecenters or private intermediaries selling airtime. The use of dedicated intermediaries such as individuals or community centers has the advantage of reducing customer management costs for telecom operators. This is one reason why Grameen Phone succeeded.

New Forms of Top-Up

Mobile payment schemes need to change to address low-income end users. Prepaid schemes extended mobile services to the masses in mature markets, but the average price of the prepaid scratch card is generally too high for low-end users. Most of these customers do not have a credit card or bank account and only a tiny sum to spend on mobile communications. Considering that low-ARPU segments range up to US$5, telecom operators should provide the widest choice of micro-top-up options, such as cheaper recharge cards (for example Nokia's US$0.5 recharge card with limited validity), or other alternatives like e-refill solutions, over-the-air payment solutions or micro vouchers. Some operators even offer airtime swapping options between nations, enabling users from one country to send credit to someone in another country. By enabling users to reload their phones without using physical debit or prepaid cards, operators reduce production and distribution costs (OPEX), and can decrease churn rate (customers switching to another network operator). Churn can be also minimised with additional services based on consumer demand.

Case study 2: *Innovative e-refill solutions*

E-refill allows subscribers to pay the reseller of their choice for the desired amount of airtime, just like buying vouchers. Upon receipt of cash from the subscriber, the reseller sends a payment request message to the prepaid system. The subscriber's account is credited automatically, and both parties are notified that the transaction has occurred and the latest account balance is provided in a text message to the subscriber. Over-the-Air Refill (OTAR) solutions satisfy operators' key refill targets: more revenue generation, and simple and fashionable refill methods for end users. Due to the heavy cost structure of standard scratch cards, mobile operators are very keen on introducing electronic or over-the-air refill capabilities (using SMS) to their subscriber base. By suppressing the manufacturing and warehousing costs linked to physical vouchers telcos can:

- *Easily increase their margin on each top-up transaction;*
- *Increase revenue generation and subscriber numbers by launching new marketing offers;*
- *Enhance their offer by providing customers with a new and fashionable refill method.*

E-vouchers enable mobile operators to launch entry-level, prepaid packs with micro-prepaid refills dedicated to low-ARPU segments. Thanks to the e-voucher solution, some amazing marketing successes have already been achieved in emerging markets. In addition to e-voucher, person-to-person refill (P2P) transactions are a complementary tool enabling mobile operators to provide more flexibility to their customers in the way they use their own credit. This P2P facility is highly appreciated within a family or within a community, since credit can move from one account to another, for example between a father and his son, or between two friends. It allows the unblocking of prepaid users who have reached their credit limit or expiry date. It is also a way of granting small amounts of credit to relatives, or to people who cannot afford to buy a complete scratch card. Examples of telecom operators that have implemented e-refill solutions include: Smart Globe and Digitel (Philippines), Airtel (India), Exelcom (Indonesia), Comvik (Vietnam).

Providing Affordable Value-Added Services

Service Differentiation

Public authorities and regulators have a strong role to play in opening competition and decreasing import taxes and duties on equipment and services. The cost of telecom services often remains too high, and not always in line with the disposable income of many African communities. Operators should also be encouraged to offer different tariffs depending on the level of service, customer profile and consumption patterns. In effect, a distinction should be made between urban and rural customers, to provide the right service to the right segment. Low-income segments will require fewer or more basic services, and their lower level of consumption could justify a difference in pricing. Tariff plans are fundamental to encouraging consumption, and to retain customers through promotions, discounts, and special offers. Group and family plans for both prepaid and postpaid are a strategic way to increase loyalty. Finally, the gap between pre-paid and post-paid communication rates should also diminish, as the prepaid scheme has become the reference model in developing countries, partly due to poor access and low banking service penetration, especially in rural areas.

Value-Added Services and Applications

A good understanding of the needs of rural customer can help in the design of value-added services and applications. This could take the form of applications developed by service providers which telecom operators could resell: e-banking, e-health, e-trade, market price information applications, etc. There are also core network-enabled features adapted to rural use: Push to Talk over cellular, audio messaging (alternative to SMS), Fixed Cellular Subscriber (FCS), emergency warnings through SMS, etc. These services should be offered with local content, and take into account the constraint of illiteracy. With the region's limited fixed-line penetration effectively limiting Internet access via traditional access methods, mobile technology now has the opportunity to drive Internet use.

Time for Wireless Data

So-called 2nd-generation services such as WAP (Wireless Application Protocol) or SMS (Short messaging service) are gaining ground. Innovative, region-specific applications have also helped drive SMS and WAP usage: Mobile banking in Nigeria for example or providing election results in Kenya. Interest in these applications indicates a broader level of demand for data services. GPRS (General Packet Radio Services) or EDGE (Enhanced Data Rates for GSM Evolution) are being launched in an increasing number of the region's markets, with a number of other networks now GPRS/EDGE ready. With the capacity to provide higher-speed Internet access, GPRS/EDGE can provide an initial solution to the lack of Internet access in emerging markets.

Community-Based Applications

There are also community-based applications, which could be provided through community intermediaries, aiming at improving public services and reducing isolation. There are numerous cases of applications in the fields of healthcare or education that help reduce the so called digital divide between urban and rural areas. Digital radiology for example, even if it requires heavy investment in hardware, brings economies of scale in terms of equipment and staff, while improving quality of service. It also enables public health targets to be reached, offering remote diagnosis services for isolated people in rural areas.

KEY CHALLENGES FOR SERVICE PROVIDERS

There are different parameters that make rural and remote areas less attractive than urban ones for communications operators, mainly as a result of costs and revenues. These can be categorised into a number of dimensions. Each of these dimensions is discussed next.

Dimension 1: Network Coverage and Capacity

Even though there has been lot of investment made in the telecom infrastructure over the last five years there still is tremendous need for network capacity, subscribers are queuing up for both mobile and fixed connections. For ease of discussion and understanding let us take the same two countries as examples. In Nigeria and Algeria, despite of the enormous growth in active subscriber counts over last 3 years, there still is a serious lack of network capacity. The conservative subscriber growth forecast in both countries is expected to be in the range 40 to 50% CAGR. The challenge is for the operators to build networks and operations frameworks which are capable of managing this sustainable growth.

Dimension 2: Quality of Service

Maintaining quality of service across the region is seen as a fundamental challenge. The down times and speech quality are very poor and need to be addressed. The networks across the continent need to be optimised and utilised to the maximum, also making sure the quality parameters are maintained. There has to be a shift from CAPEX to OPEX to make sure the investments made in CAPEX are used to the maximum.

Dimension 3: Churn and Revenue Uncertainty

Largely owing to a predominantly prepaid subscriber base, the churn rates are higher compared to more developed markets and the main reason for this is the lack of churn management strategies. Average churn rates across the continent are exceeding 20%. The African market is predominantly prepaid and the share of the prepaid base ranges between 93–98% across different operators. In most of the operators across the continent, prepaid tariffs account for 75–85% of total revenue. However, the prepaid billing environment has a downside. Prepaid platforms notably have difficulty of controlling the customer churn and building revenue certainty.

The challenge for the operators is to focus on working against this by deploying service management strategies and tariff packages that provide the flexibility of prepaid along with better customer management and increased revenue certainty.

Dimension 4: Pricing Innovation

Telecom services pricing and tariffs in the African continent used to be fairly simple, with mobile operators differentiating only between prepaid and post-paid. However, as competition is intensifying and penetration levels are increasing, the pricing packages must grow further in sophistication. In line with this, the focus on customer care and evolution in billing platforms is helping to increase the average usage and improving the customer retention in the markets where 20–25% of the customer base has been leaving the operators annually.

Operators need to emphasise on the following attributes while rolling out new service management and revenue growth strategies 1) Customisation; the ability of a service package to meet the need of the subscriber, to meet usage patterns and affordability level of the subscriber 2) Simplicity; the ability of the package to offer clear and easily understood pricing, allowing ease of evaluation 3) Cost Saving; value for money in a market where mobile services are considered very expensive.

Dimension 5: Operational Costs

Telecom services operations are run very inefficiently compared to the developing operators in other parts of the world. The operations cost vis-à-vis the quality levels ratios are poor. Despite the poor quality levels, the average cost of management is high and hence the OPEX is high.

The fundamental reasons for the high OPEX are the following:

Network capacity and coverage: Very high costs for operators to reach people in remote areas and high costs of rolling out telecom networks to the remote areas where the number of user and traffic is very low. Hence return on investment is too slow.

Lack of local competence: Dependency for skills on the expatriate workers because of the lack of telecom and mobile operations experience in the region. This adds lot of load to the operational expenditure (OPEX). The broader components of the OPEX are network operations and maintenance (amounts to 50–60%), G & A (Including Staff) (10–15%), sales & marketing (5–10%), service and customer management (10–15%), interconnect costs (10–15%). So, in order to address the problem of high OPEX, operators need to find innovative ways of controlling the high cost components of overall OPEX e.g. network operations and maintenance.

Inefficient operations framework: The lack of the proven process frameworks and operational tools results in reactive maintenance which, in turn, worsens rates of network downtime, customer satisfaction, quality of service etc. and increases the cost of retention and the overall cost of management.

Staff costs versus productivity are high and the number of employees per subscriber varies amongst African operators by as much as 300%.

High customer churn rates: Due to the predominantly prepaid customer base, African churn rates are much higher than the average in developing markets. However, it is difficult to explain the actual average churn rates across different markets of the continent. If I take an example from sub-Saharan Africa, South Africa is experiencing the continent's highest churn rate, ranging between 30–40% among the three mobile operators. But, if we take a rough estimate, the average churn across the continent should be in the range of 20–25%. Putting this in commercial perspective, operators on average loose 1 out of four subscribers acquired. This directly affects their cash flow. Operators are trying hard to reduce these churn rates by implementing retention strategies and this has direct impact on the cost of management.

Dimension 6: Inadequate Financing Solutions

The main obstacles to increasing the penetration levels are high start up costs e.g. 1) cost of handset and 2) the operator's ability to reach and service the end user, reach and awareness generation through extensive sales channels, adequate promotions, service through adequate coverage, network capacity customer care and overall quality of service.

To overcome these obstacles, operators need to come up with innovative financing options and schemes e.g. subsidised handsets, deferred payment schemes, better customer care and network coverage and quality.

UNDERSTANDING THE VALUE CHAIN

Key Players Involved in the Process of ICT Development Public Bodies

The role of the public sector in the high technology content of telecom services is supplemented by regulation policies and standardisation activities. Public bodies have a very important role to play in the process of ICT development, to foster infrastructure roll-out in unserved areas and enable universal service provision. It has been proved that the creation of a separate regulation authority has a positive impact on the efficiency of the telecom sector and can accelerate network development. ICT policies and strategies should target taxation to allow telecom operators to bring tariffs down, and open markets to facilitate market entry for new private players. Since medium/high income consumers are the most lucrative markets, specific licenses and funding will need to be made available to convince investors and private players to go rural. Specific licensing for underserved regions has been implemented in South Africa and is under study in other sub-Saharan countries. Indeed, government and public administrations are themselves the first potential customers. The means to access public documents, to ease administration, to share administration information, and to establish connections between citizens and the administration are all being heavily promoted by e-government initiatives.

Operators

Existing operators (telecom and ISP), especially in developing regions such as SSA, have not yet invested much in access delivery in rural and remote areas, even if they are supposed to do so according to universal service obligations. Rural markets are often considered risky, due to high entry costs and smaller revenue opportunities (low income/low population density) compared to urban markets. However, real future growth for African telecom operators will come from low average revenue per user (ARPU)/rural markets as high-end segments (urban/high ARPU customers) start to saturate. By deploying infrastructure in rural areas, operators can increase their customer base and revenues; lower margins will be offset by higher-volume revenues.

Services and Applications Providers (SAPs)

SAPs design value-added services targeting end users. Scalability of service can be achieved if SAPs establish partnerships to capitalise on the distribution channels of telecom operators. Both service providers and operators benefit from these new forms of agreement that allow them to reach a critical mass of users. Commercial agreements between SAPs and telecom operators are not yet highly developed in Africa, but could have a positive impact on the availability and affordability of services for end users.

End Users

End users are individuals and professionals that could potentially become ICT users if they can have access to a terminal at a reasonable cost. Mobile penetration in most developing countries leaves room for major growth, and demand will grow as tariffs decrease, and new handsets appear with design and features adapted to the needs of low-income users. To accelerate uptake of ICT and its benefits (improved quality of life, access to knowledge and information), efforts should also be made, maybe by public authorities, to help people to familiarise themselves with ICT and have access to credit. In addition to the providers in the value chain listed above, other stakeholders also have a role to play.

Public Facilities

Public facilities would benefit from having access to technological innovation, which could improve the quality of public service, offset the lack of resources at remote sites, and help to connect rural and urban areas. In some cases, economies of scale (staff and equipment/material) can also be achieved by using new technologies rather than traditional methods (digital radiology for example). Giving Internet access to rural users, through collective or shared infrastructure (schools or telecenters), even if it will not bring fast profits, will have a strong impact on human capability and local economic development. ICT can be seen as a tool for achieving sustainable economic growth, enhanced public welfare, improved transparency, and social and economic stability. Public facilities could be used to run community information and promotion programmes on the use of ICT to develop human capability and raise empowerment, through cybercafés or telecentres for example.

Donors

Donors can share the risks of financing rural access, making the cost of entry less dissuasive for telecom operators. Funding could come from the private sector (tax on telecom operators) and national bodies, setting commitments in terms of universal service provision, or local players interested in boosting the attractiveness of a specific region. Part of the financing could also come from international donors: development agencies and international organisations (United Nations, World Bank, European Commission, African Development Bank etc). Financial support could directly target end users, helping to decrease terminal acquisition costs, or indirectly, meaning that the service provider would benefit from the funding. As described in the *Rural ICT Toolkit*[2] report from African Connection, a smart subsidy is an initial subsidy given to the private sector that is result-oriented, does not distort the market, and encourages cost minimisation and market growth. It helps to kick-start a project or service delivery, using contracts that tie payments to the benefits actually delivered to target beneficiaries. In addition, outside funding, possibly public, would be justified when:

- A project could make a profit but is considered only marginally viable and marginally attractive to investors in the short term, and/or low priority, without the incentive of a rural subsidy;
- A project will be commercially viable if high start-up costs (capital-intensive infrastructure) are partly funded.

Creating Value for Communications Development in Underserved Areas

How can communication tools like the Internet and telephone contribute to the local development of communities that are often disadvantaged by the lack of even basic facilities such as drinking water, roads or electricity? Why, in this situation, should ICT investment be not only useful or a priority, but even economically realistic? Various case studies like Manobi and Pésinet[3] show how it is possible to develop *proximity* services, that is, services that meet the basic needs of the local economic and social organisations and the poorest people. Such services must be defined locally, taking into account people's way of life, real needs and incomes. The model presented in the next figure sets out ways in which ICT could contribute to a lasting, integrated development process. It is mainly based on offering high local added-value proximity services, unlike the standard use of the Internet in the industrialised countries. Here the term

[2] http://www.infodev.org/projects/telecommunications/351africa/RuralICT/Toolkit.pdf last accessed 12 February 2008.
[3] Marine, S. and Blanchard, J-M. Strategy White Paper. Bridging the digital divide: an opportunity for growth for the 21st century. *Alcatel Telecommunications Review*. 3rd quarter 2004.

transparent market refers to local trade, such as the activities of producers and fishermen using Manobi services for example, while the term *health-care* highlights the benefits of service like Pésinet, offered by Senegal's Saint Louis hospital to those in the poorest areas. This development model shows how both local players and local residents can achieve a genuine leap forward economically, politically and socially, based on two converging virtuous circles.

The model shows that:

- Lack of infrastructure and illiteracy are two prime causes of sustained poverty; access to information to take care of oneself, feed oneself, communicate with peers, develop projects, etc. can be a lifeline for isolated communities.
- Mobile is unquestionably the most realistic investment in communication, because of the quicker return on investment compared with alternative, costly infrastructures. The Internet cannot replace the roads that are so sadly lacking, but suitable Internet services will make it possible to make better use of what few means of transport are available.
- Economically, mobile infrastructure will help to create local, more transparent marketing channels, so limiting speculation and the risk of artificial shortages and improving the distribution of margins between the various links in the value chain of each sector, from producer through to consumer. Time and money saved in this way can be ploughed back into productive new activities, helping to boost the local economy and leading to the creation of jobs. This will, in turn, justify more communication resources, and so on. This is the first virtuous circle.
- The second virtuous circle is of a social and political nature, in which ICT can be used as a tool to support the implementation of healthcare initiatives in which information campaigns are so important. In the areas of education and how society works, the Internet has the potential to improve communication between public authorities and local people, as well as between central authorities and local authorities. It will facilitate greater transparency in how institutions are run, moving towards the objective of good governance, and offsetting the lack of transportation infrastructure and local government presence.

Infrastructure Implementation Strategy Based on Usage

In developing countries, the main barrier to setting up communication infrastructures is the lack of available investment. This problem is even more critical in rural areas that are still very poorly served. Thus, an approach is required based primarily on usage and services and in which the technology is not considered as an end in itself, but more as a tool.

There is enormous potential in this area, comprising a multitude of initiatives based on individual competences or small creative and dynamic organisations that can develop new proximity services. Such initiatives warrant support and mentoring. To this end, the public authorities will have a key role in creating conditions that favor the lasting emergence of such potential. The second step is to nurture the most promising initiatives and set up larger scale trials or pilot projects. This type of project could be usefully financed by public start-up funds, possibly in partnership with private financing through public-private partnership (PPP) schemes.

A pilot project's essential goal must be to study the economic viability of the proposed service platforms, if balanced business plans are to be drawn up. Lastly, when as many so-called demonstration pieces as possible have been set up through the pilot projects, potential investors (public or private) can commit to the large-scale deployment of infrastructures based on conventional cost effectiveness criteria.

The Importance of Public-private Partnerships

Diverse models can be implemented depending on national regulation, local community access objectives and local conditions. Projects include not only operators, service providers, wholesale operators and telecommunication suppliers, but also new actors, such as civil works and construction companies, utilities and financial institutions. Public-private partnerships can generate new types of consortia, in which these actors may become partners or investors in the deployment and operation of regional or local broadband networks, for example. Private and public partners can share initial capital expenditures (infrastructure, civil works) and future revenues.

An Appropriate Regulatory Framework: Universal Access and Rural Licensing

In countries with large imbalances in communications development among different regions, regional licensing enables governments to target underserved areas with specialised licenses or more favorable treatment of rural areas. Potential licensees in these areas may be attracted, for example, by exclusive licenses. Rural licenses can also be bundled with licenses to provide services in more lucrative markets. For example, underserved counties in Uganda are packaged into three separate Universal Access Regions for licensing purposes.

Each Universal Access Region bundles together a mix of counties with different levels of market potential. Bundling can also combine rural licenses with rights to offer more profitable services such as international long-distance and cellular mobile services. In such cases however, regulators should ensure that anticompetitive cross-subsidisation does not lead to predatory pricing that drives out of business competitors that are licensed only to provide rural local services. One way to avoid this is to make such bundling available to all rural providers. By contrast, a licensing approach that authorises certain carriers to offer service only in rural areas raises a number of concerns. The ability of any such carrier to attract large-scale capital investment could be questioned. There may be doubts about the long-term sustainability of rural providers, given the lower revenues they can generate in these areas. For this reason, some countries have tried to start small, by focusing their licensing approach on encouraging small and medium-sized enterprises to enter rural markets by lowering entry barriers, compared with the process for entering urban markets.

Governments can implement some options that make entry into rural and underserved markets more attractive to small and medium-sized enterprises. They can lower rural licensing hurdles such as non-recurring fees, and large performance bonds that are normally attached to licenses for the provision of basic telecommunication services and facilities. They can also relax performance mandates designed to maintain a high quality of service, and they can reduce stringent tariff requirements. Rural licenses for small and medium operators could also be subject to lower annual licensing fees and exempted from contributions to Universal Service funds. Spectrum for the deployment of cost effective wireless broadband technologies could also be offered to rural licensees, at reduced fees or through auctions or reverse auctions, in order to encourage their deployment. In some countries, arguments have been made for imposing asymmetrical termination charges between the incumbent operators and rural licensees. Such an approach would allow rural licensees to command larger termination charges than they would have to pay to the incumbent operator.

DEVELOPING SUSTAINABLE BUSINESS MODELS FOR RURAL NETWORK OPERATORS

This chapter highlights financial innovation and cost-reduction strategies that may convince telecom operators to roll out infrastructure profitably in rural areas, and to exercise pressure on public policymakers that may have previously granted license exclusivity to one operator that is unwilling to serve rural areas. As mentioned before, telecom operators in sub-Saharan Africa have not invested much in rural areas, due to the heavy investments required and low margin opportunities compared to lucrative urban markets. However, Universal Service provision will soon be a commitment for telecom operators with financial support from dedicated resources (Universal Service Fund), and new business opportunities in urban markets have started slowing down. Therefore, it could be interesting for operators to consider rural areas that have not yet been served, and could become a mass market, if they are well understood and addressed with appropriate localised solutions. For service providers, the key consideration is how to optimise both CAPEX (capital expenditures: all costs related to initial investments) and OPEX (operational expenditures: annual cost of running the network) to boost revenues from existing and new customers and achieve a fast return on investment.

Financial Innovation for Operators

There are different financial schemes and innovative payment options for operators depending on their needs, and according to their financial capabilities and strategy. Telecom equipment and infrastructure suppliers are able to offer a wide array of customised solutions that make financial sense for communications operators and service providers.

Innovative Vendor/Operator Partnerships

Innovative vendor/operator partnerships offer the possibility to develop the network following traffic growth or customer base increases, on a *pay as you grow* basis. The operator in this case will pay for the used capacity. Telecom suppliers and operators can also choose to share revenues from running services, either for specific applications, or for all revenues and commercial risks.

Innovative Financing Options

Innovative payment options are also available that allow operators to focus on their core business: the delivery of competitive subscriber services, customer care, marketing, branding and development. This could be achieved with zero CAPEX; for example, the operator can decide to retain ownership of the network and outsource running tasks to an outside contractor: hardware operation, software management, network optimisation, personnel training, etc. Ownership can alternatively be kept by the telecom equipment supplier (turnkey project), with the operator purchasing level of coverage, capacity and quality without being involved in day-to-day operational activities. The operator can in this case negotiate an option to buy the equipment. Deferred payment options can also be agreed to support the operator in the set-up phase.

Total Cost of Ownership (TCO) Reduction

Delivery of access to rural areas should be cost efficient, and maximise use of the operator's three major assets: subscriber base, base station sites and spectrum (licensed and unlicensed). Equipment suppliers are now able to provide solutions that fit all types of areas and their associated constraints. Cost efficient solutions aim to lower the total cost of ownership for telecom operators, making the extension of the network to rural areas possible and profitable if the customer base is sufficient. The need for low-CAPEX products that enable an operator to provide competitive services (reduced CAPEX is essential for financial reasons) is at the heart of a significant shift in technology focus. The ability to increase network capacity in the future therefore depends almost entirely on the availability of low-cost equipment compatible with a low Average Revenue Per User (ARPU). Network solutions based on a full product portfolio are shaping the future evolution of these markets: narrowband and broadband switching and highly-scalable access nodes must be available at a very competitive price.

Cost Reduction Strategies

There are multiple ways for communications operators and service providers to reduce costs affecting both CAPEX and OPEX, in other words achieving reduced network investments and operational expenses. CAPEX efficiency can be achieved by preserving the initial investment, for example re-using existing sites when a mobile network is extended or upgraded. Network OPEX can be optimised with new core network features such as remote upgrade via software, reducing maintenance costs. Power consumption is another opportunity for cost reduction. Indeed, since power requirements are a critical consideration in technology selection for rural areas, devices should always be designed for minimal power consumption.

Tailored, Cost-Efficient Solutions for Rural Areas

Improvements to core network, access and backhauling solutions are constantly being developed through R&D to reduce costs. Adaptations to suit rural applications result in the design of cost efficient solutions with enhanced capacity and coverage.

Shared Network Infrastructure

Network infrastructure costs can also be shared among several operators interested in addressing different segments or services within the same region. Existing operators could also resell part of their capacities to alternative operators (see Mobile Virtual Network Operators).

Case Study 3: *Mali*

This section will draw on our previous conclusions to show how ICT deployment in low-income areas (rural and remote) could be profitable if markets are well understood and correctly addressed. As expressed earlier, this is a question of the availability and affordability of the services for rural users. GSM and WiMAX have been chosen as access technologies, because of their cost efficiency and capabilities in rural environments, allowing the fast and efficient roll-out of voice and data services, with sufficient bandwidth (in the case of WiMAX) to support enhanced applications for collective use. As discussed earlier, GSM best suits voice and data applications (at low bit rates) and is widely deployed in Africa. WiMAX is a growing standard offering broadband Internet access, and is an alternative to wired technologies (DSL, cable, fiber), where no copper infrastructure exists. Calculations and assumptions are based on Ericsson's documentation on radio access solutions for low-ARPU areas. We took the case of Mali, which is one of the most challenging countries in SSA, considering its current levels of GDP per capita and mobile penetration, especially in low-density areas. Mali remains one of the poorest countries in the world, with 65% of its land area desert or semi-desert and with a highly unequal distribution of income. About 10% of the population is nomadic and some 80% of the labour force is engaged in farming and fishing.

Some Key Figures on Mali

- Area 1 241 000 km2
- Population: 11.5 million (2004)
- GDP (Purchasing Power Parity): US$14.18 billion (2007 est.)
- GDP per capita (PPP): US$900 (2004 est.)
- Literacy: total population 46.4% (+15 years old), male 53.5%
- Population below poverty line: 64% average, 30% of total population living in urban areas, 70% of total population living in rural areas (2001 est.)

ICT at a Glance

- Radio broadcast stations: AM 1, FM 28, shortwave 1 (seven frequencies and five transmiltes and relays China Radio International) (2001)
- Television broadcast stations: 1 (plus repeaters) (2001)
- Internet country code: .ml
- Internet hosts: 28 (2007)
- Internet users: 70 000 (2006)[4]

GSM Business Case for Rural Coverage

We first considered a business case focusing on two segments that have not been addressed: (1) low-urban and (2) low-density areas. From the mobile operator's perspective, these areas can be considered as low-ARPU segments. These two segments represent about 80% of the total population.

Market Assumptions

The mobile penetration assumptions are based on a Credit Suisse model linking adoption rate with the percentage of mobile expenses (ARPU + handset) by GDP/capita, as follows:

- Three users sharing one mobile handset/line;
- ARPU (i.e. per line): US$4 per month (low-urban) and US$3 per month (low-density), with 80% of revenues coming from voice and 20% from data (mainly SMS);
- Prepaid 100%, churn rate (annual) 10%;
- One single radio network deployed, resulting in one network provider (maybe incumbent), offering possible access to Mobile Virtual Network Operators (MVNOs);
- Handset average selling price of US$40 in 2006, decreasing by 5% annually.

Overview of Mobile Penetration Evolution for the Two Segments

CAGR refers to compounded annual growth rate. Mobile penetration is expected to grow from 0% to about 20% in low-urban areas, and from 0% to about 15% in low-density areas. As mentioned, mobile penetration will be influenced by growing GDP/capita as well as by decreasing handset costs. Other parameters can also contribute to

---

[4]CIA Worldfactbook. https://www.cia.gov/library/publications/the-world-factbook/geos/ml.html. Last accessed 12 February 2008.

faster mobile adoption, like decreasing tariffs and targeted value added services and applications.

Overview of Revenues and Margins

After CAPEX and OPEX assumptions, not detailed here, we see a positive EBITDA contribution in years 2–3.

Overview of Free Cash Flows

After an initial investment of US$141m, free cash flow break-even is reached between years two and three, with an average positive contribution of US$50m per year over the period. Coverage extension of low-urban & low-density areas generates an NPV contribution of US$11.88m over 15 years.

Investment payback occurs in year seven.

Concluding Remarks

Overall payback is reached in seven years, which is quite long when compared to classic projects in urban environments. Internal Rate of Return (IRR) is also far lower than usual telecom projects, which average 25–30%. However, it should be observed that the low-density segment complicates the business case, due to insufficient subscriber density per BTS site (less than 500 subscribers over the period, vs. 1000 usually needed to reach profitability). The business case could be improved with a focus on areas having more than 10 inhabitants/km. In addition, as described in the introduction of the case study, Mali is one of the poorest countries of SSA. The business case focusing on 'low ARPU' segments could be improved with countries having higher GDP/capita. Thus, to make the business case in some rural and remote regions attractive for telecom operators, it would be relevant to consider reasonable public intervention to decrease initial CAPEX.

BEST PRACTICES THAT LOWER TCO

There are several types of activities available for the cost reduction of both OPEX and CAPEX.

Improvement of Processes and Routines

By changing the ways of working and increasing speed and quality of work activities, lead times and total number of activities can be minimised. This will reduce the overall manpower needed to run the operations. These types of improvements are primarily interesting for network operations, CRM and billing activities.

Improvement of System Performance

Outsourcing can be used to complement or be part of process and routines improvements. This will further enhance headcount reductions and ensure maximised economies of scale and cost control is achieved. In addition to outsourcing of operational activities, assets can also be included e.g. spare parts.

Outsourcing

Outsourcing can be used to complement or be part of process and routines improvements. This will further enhance headcount reductions and ensure maximised economies of scale and cost control is achieved. In addition to outsourcing of operational activities, assets can also be included e.g. spare parts.

Infrastructure Sharing

By sharing network infrastructure with one or more other operators significant reductions of both investments and associated operational costs can be achieved. Obviously these type of savings are primarily related to network and IT system infrastructure.

Partnership for Cost Sharing

Besides from sharing infrastructure operators can also share other costs with competitors and or partners. This could for instance be sharing of marketing and distribution costs by working together with brands in other industries.

Market Offer Improvements/Changes

By changing the way to communications operators and service providers market and sell their services they can cope with the high costs for marketing, sales and distribution. Activities include new and innovative ways to subsidise terminals, sliming of the product service portfolio etc. Quite obviously there are other best practices that we describe below:

Site Sharing

Description

The purpose of co-siting is to achieve advantages in time to market and cost reduction when building a network. A typical arrangement involves sharing of antenna space, equipment housing, power supply equipment, sharing of the site premises equipment including antenna, transmission and transmission management resources. Cost savings are derived through reduced investment and reduced operational costs.

Key Assumptions

- OPEX is 60% of revenues, depreciation is 15% of revenues.
- Network operations is 20% of OPEX and 15% of revenue, site costs are 43% of depreciation.
- 25% of the network operations costs are assumed to be associated with the sites, this includes site rental facilities management, utilities etc., these costs are assumed to be reduced by 50% (i.e. the sites are shared between two operators building for coverage not capacity). Accordingly the investments in sites are assumed to be reduced by 50%.

Resulting Savings

Resulting savings are shown in Table 6.1.

Shared Radio Access Network

Description

The purpose of shared radio access networks is to achieve advantages in cost reduction when building a radio network. A typical arrangement involves shared base transmission stations and radio network controllers. The operators control their radio network by operating their own

Table 6.1

Area	Savings % of OPEX	Savings % in dep.	Savings % of revenue	Savings in Base Case (mEUR)
Site investment	N/A	21.5%	3.2%	18.2
Site operations	2.5%	N/A	1.5%	8.5
Total	2.5%	21.5%	4.7%	26.7

transceivers within the BTS and by using their licensed frequencies. In this arrangement, cost savings are realised in a number of areas: equipment installation and commissioning, radio network planning, transmission etc. The cost of manpower affects OPEX and the HW CAPEX.

Key Assumptions

- OPEX is 60% of revenues, Depreciation is 15% of revenues.
- Network operations is 20% of OPEX and 15% of revenue, Site costs are 43% and radio network is 33% of depreciation.
- 25% of the network operations costs are assumed to be associated with the sites and 50% with the radio network, this includes site rental facilities management, utilities, maintenance, support, planning & optimisation radio NOC etc., these costs are assumed to be reduced by 50% (i.e. the network are shared between two operators building for coverage not capacity). Accordingly the investments in sites and radio network are assumed to be reduced by 50%.

Resulting Savings

Resulting savings are shown in Table 6.2.

Efficient Sourcing

Description

Sourcing is the identification, evaluation, negotiation and configuration of products and suppliers. Cost savings are primarily achieved by reduced prices paid for goods and services. In addition, savings may also be achieved by improving the processes and supporting systems for the procurement activities (not considered in business case as saving).

Table 6.2

Area	Savings % of OPEX	Savings % in dep.	Savings % of revenue	Savings in Base Case (mEUR)
Site investment	N/A	21.5%	3.2%	18.2
Site operations	2.5%	N/A	1.5%	8.5
Radio investment	N/A	10.9%	1.6%	9.2
Radio operations	5.0%	N/A	3.0%	17.0
Total	7.5%	32.4%	9.4%	52.9

Key Assumptions

- Depreciation is 15% of revenues.
- Strategic sourcing activities can help reduce costs for all investments made.
- Cost savings are expected to reach 15%. This is in line with benchmarks from AT Kearney amongst others.[4]

Resulting Savings

Resulting savings are shown in Table 6.3.

Improved Network Performance

Description

The purpose of improved network performance is to achieve cost efficient utilisation of the complete network. There are three major areas that area subject to network efficiency improvements: Radio, Transmission and Core. The utilisation can be achieved through the tuning of existing infrastructure or by investments in new infrastructure or support systems. Cost savings are achieved by reduced long term operational costs thanks to less maintenance and surveillance requirements.

Key Assumptions

- OPEX is 60% of revenues.
- Network operations represent 20% of OPEX and 15% of revenue.
- The improved network performance is expected to decrease O&M costs by 10% for the affected areas.
- The assumption is that the affected areas (core, transmission and radio are all included) make up 60% of the total network operations costs. Costs for leased lines, site rental etc is not included.
- The following savings and benefits have not been included:

Table 6.3

Area	Savings % of OPEX	Savings % in dep.	Savings % of revenue	Savings in Base Case (mEUR)
Efficient sourcing	N/A	16%	2.4%	13.6
Total	0.0%	16%	2.4%	13.6

[4] http://en.wikipedia.org/wiki/A.T._Kearney

Table 6.4

Area	Savings % of OPEX	Savings % in dep.	Savings % of revenue	Savings in Base Case (mEUR)
Core	0.2%	N/A	0.1%	0.7
Transmission	0.3%	N/A	0.2%	1.0
Radio	1.0%	N/A	0.6%	3.4
Total	1.5%	0.0%	0.9%	5.1

- reduction of handset subsidies due to reduced churn
- less HW and SW investment due to cost efficient utilisation of network
- Increased revenues.

Resulting Savings

Resulting savings are shown in Table 6.4.

Outsourcing of Network Operations

Description

The purpose of outsourcing of network operations is to achieve cost reduction. For the outsourcing partner, this is achieved by scale of economies and know how in terms of improved processes, systems and tools. The outsourcing of network operations cost saving pack includes outsourcing of the NOC, field maintenance, build and planning & optimisation activities. Note that the learning curve has to be taken into consideration. The savings cannot be fully realised from day one.

Key Assumptions

- OPEX is 60% of revenues.
- Network operations is 20% of OPEX and 15% of revenue.
- NOC, field maintenance, build and planning & optimisation constitutes 50% of Network operations cost. The rest is made up of leased lines, site rental, utilities, support cost etc. not affected by the outsourcing activities.
- Savings in the outsourced areas is expected to be 15%.

Resulting Savings

Resulting savings are shown in Table 6.5.

Table 6.5

Area	Savings % of OPEX	Savings % in dep.	Savings % of revenue	Savings in Base Case (mEUR)
Network Op C	0.3%	N/A	0.2%	1.0
Field Maintenance	0.8%	N/A	0.5%	2.5
Build	0.3%	N/A	0.2%	1.0
Planning & opt.	0.2%	N/A	0.1%	0.5
Total	1.5%	0.0%	0.9%	5.1

Efficient Customer Care

Description

The purpose of efficient customer care is to assist the customer in administering and providing end user services efficiently, for example: efficient end user service provisioning (self configuration systems), more efficient support systems facilitating support personnel's work, improved CRM processes reducing lead times and minimising errors and competence development.

Key Assumptions

* OPEX is 60% of revenues.
* Customer care % administration is 15% of OPEX and 9% of revenue.
* The improvements are expected to affect the full CC&A cost, i.e. 100%.
* Savings from system and process improvements are expected to generate 5% savings each (10% in total) and savings from increased customer self-management is expected to generate 5% in savings (at least initially).

Resulting Savings

Resulting savings are shown in Table 6.6.

Outsourcing of Billing

Description

The purpose of outsourced billing is to completely relieve the operator of all activities associated with the billing (however for simplicity the

Table 6.6

Area	Savings % of OPEX	Savings % in dep.	Savings % of revenue	Savings in Base Case (mEUR)
Improved CRM system	0.8%	N/A	0.5%	2.5
Improved CRM process	0.8%	N/A	0.5%	2.5
Increased customer self management	0.8%	N/A	0.5%	2.5
Total	2.3%	0.0%	1.4%	7.6

operator would still be eligible for the system investments required). The operator would send CDRs to the outsourcing partner who would take care of all activities including backend and collections activities. Cost savings would be achieved primarily through economies of scale at the outsourcing partner.

Key Assumptions

- OPEX is 60% of revenues.
- Billing is 8% of OPEX and 5% of revenue.
- The improvements are expected to affect 50% of the billing cost.
- Outsourcing savings are expected to be 20%.

Resulting Savings

Resulting savings are shown in Table 6.7.

Customer Acquisition and Retention

Description

The purpose of reduced customer acquisition and retention is to reduce the costs by introducing new ways of working with distribution, loyalty programmes, financing of terminals etc. Cost savings would be achieved through decreased churn and thus decreased costs for attracting customers.

Key Assumptions

- OPEX is 60% of revenues.
- Marketing and sales (M&S) is 16% of OPEX, advertising and promotions (A&P) is 8% of OPEX and subsidies/commissions (S&C) is 4% of OPEX.

Table 6.7

Area	Savings % of OPEX	Savings % in dep.	Savings % of revenue	Savings in Base Case (mEUR)
Backend	0.6%	N/A	0.3%	1.9
Collections	0.6%	N/A	0.3%	1.9
Total	1.1%	0.0%	0.7%	3.8

Table 6.8

Area	Savings % of OPEX	Savings % in dep.	Savings % of revenue	Savings in Base Case (mEUR)
M&S	0.8%	N/A	0.5%	2.7
A&P	0.4%	N/A	0.2%	1.4
S&C	0.8%	N/A	0.5%	2.7
Total	2.0%	0.0%	1.2%	6.8

- It is assumed that 50% of M&S and A&P is attributed directly to customer acquisition and 100% of S&C. The M&S costs are assumed to include business/service development costs and the A&P is assumed to include substantial branding and general advertising costs, these are therefore not included in CA&R.
- It is assumed that the M&S and A&P is reduced by 10% and the S&C is reduced by 20%.

Resulting Saving

Resulting savings are shown in Table 6.8.
Other practices that provide substantial saving in Opex and Capex.

- **Network sharing**
  In this practice, the infrastructure financial risks are spread out to many stakeholders. Entry hurdles for the telecom operators are lowered, as well as the total cost of ownership. Operators can focus on subscriber acquisition, customer care and service portfolio. To accommodate this, a special purpose company is established (SPC) that will own the infrastructure and a license to run the traffic. In this way the operators will pay for the capacity they are using and the SPC will outsource all operation and maintenance to the supplier.
- **Capacity growth**
  The capacity growth solution maximises the capacity on a site. In a spectrum constrained network (due to lack of foresight or simply bad planning), the capacity per site can only be increased to a certain

threshold (or there will be frequency interference). By using the frequency load planning (FLP) software (as part of the capacity growth solution), the capacity is increased beyond this threshold.

- **Coverage expansion**
  The coverage expansion solution increases the cell range of a site. As a result, fewer sites are needed. This could result in 30% fewer sites required with significantly lower power consumption. This model could be used where the operator does not want to share infrastructure and the topology is favorable.
- **Bio Coverage and remote power**
  An environmental friendly solution in areas where there is no power grid. A local network can be built powered with biofuel using distributed power and provides local job opportunities for farmers. This is a good way to involve local communities in the operational phase of a mobile communication project. Bio fuel can also be produced at lower cost than traditional petro-diesel.
- **Abis/IP and local switching**
  In order to cater to subscribers in very remote areas, blocked by jungle or mountains from urban centres, using satellite to transmit data from the base station back to the base station controller might be the only feasible solution. Installing microwave hubs would be costly and time-consuming (imagine constructing microwave towers in the middle of dense jungle). At present, satellite transmission costs are very high. However, by using local switching technology (where calls within the local vicinity are not routed via satellite, but from one base station to another), costs can be reduced by 50% (if local calls comprise 50% of total calls). In addition, the remaining traffic that will require transmission can be further reduced by using packaged switching technology.

All markets are special and local conditions vary. Therefore, the business models described may need to be adapted or even combined. The way forward could be to work in a partnership between supplier and operators.

BUSINESS MODELS THAT ARE ENABLING MASS MARKETS

The shift in strategic thinking and the growth outlook offer a significant opportunity for network operators and/or service providers and communication equipment suppliers. Regulators and governments need to play a catalytic role to encourage the models that can directly influence the challenges and weaknesses described above. All the stakeholders in the telecommunication industry need to join forces to address these challenges and reap the potential benefits of overcoming them.

The mechanism by which a company intends to create revenues and profit constitutes a business model. Both strategy and implementation are included to describe how a company intends to serve its customers. Furthermore, a business model needs to be adapted to the market it serves in order to be successful. As presented above, over half the world's population is more or less cut out of from voice and data communication, mainly in poor rural areas. To include these people in the communication era, all parties involved need to strive towards new sustainable business solutions.

The business models that carried the global subscriber base from 500 million to 2 billion were dramatically different from those used to reach the 3 billion mark, the result of a dramatic evolution in technology, regulation and business model innovation. The same mix will have to be at play again if mobile subscribers have to reach 5 billion or more. In other words, the business models that took the mobile industry to 3 billion subscribers in 2007 are not the same business models that will take the industry to 5 billion or above.

Regulation in the emerging markets is becoming more market oriented, leveling the ground and making it easier for the commercial market forces to come into play. Telecom technology is developing at a rapid pace, driving infrastructure costs down. But, business model development is lagging. Business model development involves other criteria and will require different skills for success. It involves innovation in financial modeling, a changed view of organisational structure. It could also impact the go-to-market models that corporations use today. Companies must open up for a broader partnership strategy and be prepared to work across business segments.

One example of this is the new environmental focus on the importance of biofuel. In order for the telecom industry to provide mobile communication to rural communities the issue of electrification must be solved. Telecom suppliers could/should partner with those that understand how to grow the right crops and know how to process those crops to producing produce biofuel in a sustainable way. Partnerships also need to be built with local farmer communities in order incorporate local entrepreneurs in the process. To build affordable and sustainable communication solutions, companies need to incorporate the whole value chain, also outside their own area of expertise.

There are many reasons not to enter rural areas in developing countries such as:

• High financial risk
• Cost structure (OPEX) not adapted to the low ARPU market.

- High Security risks
- Low volume of business
- Low ability to pay for services
- No power distribution
- Regulation not favorable.

The above list can probably be made much longer. However, there are also many compelling reasons to enter rural areas such as the following:

- In many developing countries Africa, India, China etc. over 60% of the population is living in rural areas.
- A mobile phone would be a paradigm shift for people in developing countries.
- Studies show that in areas where poor people obtained mobile phones, the consumption of soft drinks, cigarettes, alcohol etc. went down in favour for prepaid cards.
- By providing rural people with communication, the traffic from the already covered areas (cities) will increase heavily. Most of the people in the cities come from rural areas initially.
- Mobile communication is one basic infrastructure that will encourage many other industries to invest, thereby increasing the business development further.
- An increase in teledensity is directly correlated to an increase in GDP of a county, state and/or country.

The way to successfully address the needs of the rural poor and bring the phone to people and not the other way around is to remove the entry hurdles for the telecom operators. Most business decisions people make are based on well-known business models (traditional models). Using a traditional business model to address the issues above will be very difficult. These models are based on the culture and business climate of the developed world and will be difficult to use in the developing world where the criteria are very different. Therefore, we need to develop new business models that are adapted to the specific criteria of the developing world and specifically addressing the needs of poor people in these emerging markets. Furthermore, a business model needs to be adapted to the market it serves in order to be successful (also discussed above).

Such a model should:

- Be commercially driven for sustainability.
- Share infrastructure between two or more operators.
- Share risks between several stakeholders, enabled through a neutral infrastructure or special purpose company (SPC).
- Lower the telecom operators' need for CAPEX investments.
- Lower the OPEX cost through outsourcing.

- Add value to the subscribers/consumers through locally adapted applications.
- Stimulate high traffic growth through lower tariffs, enabled through shared infrastructure, low cost site solutions and technology adapted for rural specific criteria.
- Provide site solutions with substantially lower cost.
- Provide both voice and data (GSM and EDGE technology).
- Stimulate local market business growth.

Case Study 4: *India paving a road for mass market*

Introduction

India is a perfect combination of industry innovation, adoption of low cost model, right partnerships and sensible regulation. With the lowest tariffs in the world, India is the master of low-cost operations. Low tariffs are generating mass market appeal. At present, India is experiencing the fastest mobile subscriber growth of any market, soon to reach 200 million, growing at a rate of over 7 million subscribers a month.

Role of Regulator

The Indian regulator is working with all players in the mobile market, operators, suppliers and authorities with the objective of establishing and sustaining a good mobile infrastructure and enabling affordable mobile communications to the masses in India. Recent reports indicate that the Indian government aims to expand teledensity to more than 500 million users by the end of 2010 – around one in two of the population.

The timing is good. The vast majority of India's people live in rural areas and at the beginning of 2007 only about 60% of the population was covered by a mobile network, a fact acknowledged by the government which has declared that expanding population coverage further is a priority.

The government and changing regulatory environment are catalysing the fast adoption of wireless technologies in India. In fact, the Indian mobile market is approaching a sharp upturn as favourable demographics, strong economic growth and cheaper handsets position the country to deliver strong results in the next several years, although again, that may depend not just on how the market evolves, but whether the regulatory environment evolves further to facilitate it. To reach the ambition of achieving 50% mobile penetration in next three years, the regulator has introduced the following key initiatives:

- **Universal Service Obligation**
  Under this initiative 5% of all revenues from operator and license fee is made available to private operators that want to expand their rural networks and coverage. The fund currently is estimated to be over US$1bn.
- **Network Sharing**
  Government and regulator have introduced a system of mandatory sharing of base stations towers. There are over 8000 sites around the country, all subject to an individual tendering process with competing companies. Every new rural tower has to be shared by three operators. This cuts CAPEX dramatically and reduces OPEX for running these sites and in the end, reduces the total cost of ownership for all the operators. Also, to limit numbers of towers in the urban areas, 6 to 7 GSM and CDMA operators share each tower, achieving the same objective as in the rural area lowest TCO.
- **Plants are used to fuel networks**
  One of the clear obstacles when building rural coverage is the lack of supporting infrastructures: electricity, for example. In many rural areas, mobile connectivity is being achieved before electricity infrastructure. Even where there is electricity, it is often very unstable. To counteract this issue operators such as IDEA together with GSM Association and Ericsson have started a bio-fuel project. This project ensures that by using plants, electricity is generated locally in the remote rural areas, avoiding need for transportation and, if done at a commercial scale, will allow access to the market which would not be economically viable otherwise.
- **From voice to enriched communications**
  With globalisation of data networks such as GPRS, EDGE etc., a number of innovative data services projects are being developed to meet the desires and needs of the mass market in developing countries. Joint projects are being initiated with the banking industry to enable m-commerce applications. This is revolutionising the banking industry, enabling access to modern banking for the masses.
- **Tariffs and Operator Economics**
  Despite tariff levels as low as US$ 0.01 per minute, Indian mobile operators are, in general, coping and businesses are healthy. Despite low ARPUs (an average blended ARPU of US$7–8 and a marginal ARPU of US$3–4), operators are showing healthy EBITDA levels. The question is: how do they manage to do this? After talking to some c-level executives in India, it was apparent that they all manage to bring down the cost of production by embracing innovative business models. They have managed to eliminate all inef-

ficiencies in their operations simply by outsourcing all non-core elements of their operations such as IT, Network etc. The business model is driven with by a logic called low-cost high volume, in other words profitability is driven by economies of scale.

Some of the successful business models being used are outsourcing, manage capacity, establish operate and transfer hosting, and partnership frame models etc. These are the models that have already been tested and proven in the other market places, both developed and developing.

- **Operations Outsourcing Model**
  This business model has been very successful in other industries such as airlines and is now being successfully deployed in telecom industries with phenomenal cost saving results. Operators outsource all the non-core elements of the business for example managing technology build, operating and maintaining networks, operation and maintenance of IT/IS networks, while focusing on marketing, customer services and management.

The outsourcing business model is depicted in the figure 6.1 below:

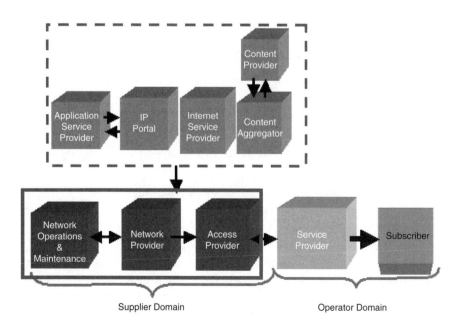

Figure 6.1 Outsourcing business model

The following are the key reasons of why operators in operators in Africa should consider the outsourcing business model:

- *Financial Pressures*
  - ARPU levels are falling as operators are heading for subscriber growth and higher penetration levels; operators can compensate this fall with new data services but this requires new investments on VAS development and marketing.
  - Competition is increasing, often from non-traditional sources; MVNOs, global telcos and resurgent incumbents.
  - High OPEX; the amount of money spent on acquiring and retaining subscribers.

Operational Inefficiencies

Staff costs versus productivity are high and the number of employees per subscriber varies amongst communications operators operating in developing countries by as much as 300%. A number of the key savings can be achieved in each of the following areas:

- Elimination of functional overlap
- Optimal utilisation of experts
- Competence management to deliver economies of scale
- Process improvement

Changing Business Operation

- Telecom services industry is changing; the customer is increasingly the driver
- Need for value-added services to retain users and grow spend
- Pressure to increase voice traffic and to grow data traffic

The realisation that the operators are not equipped for changing times, which calls for

- Application management, e.g. for the mobile Internet
- Content and partner management
- Constant network optimisation.

Figure 6.2 shows the outsourcing impact model. The outsourcing supplier's ability to absorb workload fluctuations through the scale of its operations will create greater business flexibility for the operator enabling it to respond quickly to demand variations and market changes. The introduction of output based pricing for the services offered will enable the operator to lower their fixed costs in favour of a greater variable cost element.

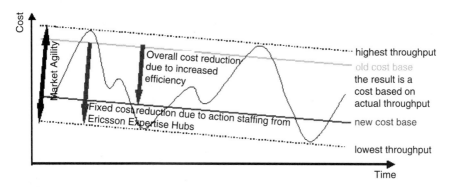

Figure 6.2 Impact of the outsourcing business model

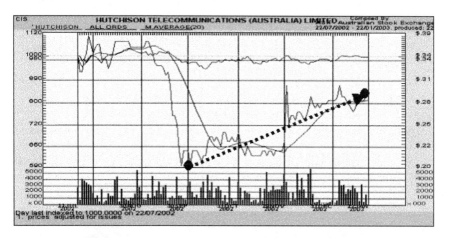

Figure 6.3 Impact on Hutchison Australia's stock value just after signing the Outsourcing Contract with Ericsson; E2E Telecom supplier

Figure 6.3 provides an example from Australia that shows the impact on the Hutchisons Australia's stock price when the operator adopted the outsourcing business model.

The key benefits to found in adopting the outsourcing business model are some of the following:

- Provides possibility for operators to focus completely on revenue generating activities
- Outsourcing will provide flexibility of predictable costs
- Will enable competence on demand
- Will give access to best practice and synergies
- Provide direct savings on operator's OPEX

- Reduces financial and operational risks
- Guarantees performance by controlled process management
- Maintains control of operator's network
- Guarantees service levels and key performance indicators
- Provides competitive pricing and benchmarking

Managed Capacity Model

Figure 6.4 shows the managed capacity business model. Managed capacity services offers operators a financially efficient way to establish a fully operational GSM network where the telco buys infrastructure capacity according to it's business needs and purchases services for network and service management operations according to network performance and services and product roll-out objectives. In this model, the operator buys the network capacity from an external supplier and, in order to manage the capacity efficiently, the supplier also takes on the network operations.

Managed capacity refers to offerings of infrastructure and long term managed services, where the fees are more linked to the operator's cash flow (the development of his business) through risk sharing.

The business concept improves and frees up the operator's cash flow to be used for other purposes. This model is particularly interesting for those government and operators that have problems of attaining funds

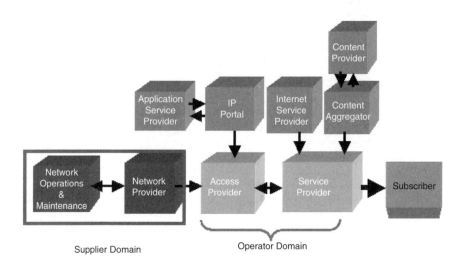

Figure 6.4 Managed capacity business model

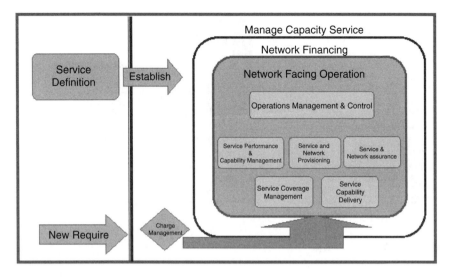

Figure 6.5 Managed capacity structure and deliverables

for expansion desired by the growth on side and lack of competence and operations experience in running the telecom networks on the other.

The value creation from the managed capacity business model is built upon with the following benefits:

- Off Balance sheet treatment of network assets
- Substantial cash flow improvement compared to traditional purchase of equipment and implementation services
- Considerably reduced pay back time and return of investment
- Innovative pricing models where the supplier shares the risk with the operator
- Proven and tested operations frameworks made available from day one
- Low OPEX and establish cost synergies
- Economies of scale through shared operations
- Best in class operation

Establish Operate and Transfer Model

Figure 6.6 shows the mechanics of the establish, operate and transfer (EOT) model. The EOT service is aimed at operators entering a new market, migrating between technologies or facing a major expansion.

The service targets network planning, design and performance optimisation, service and network assurance and service provisioning. The

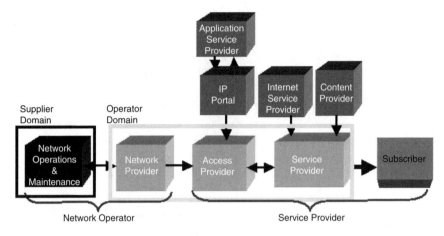

Figure 6.6 EOT model

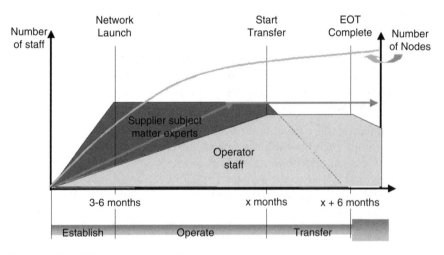

Figure 6.7 EOT model operative structure

purpose is to establish a successful operation within these areas in a smooth and efficient way and to ensure a fast time to market. The supplier takes responsibility for the operation of the network for an agreed time frame, securing optimal network performance according to defined and agreed targets.

Figure 6.7 describes the EOT model operative structure as promoted by Ericsson, one of the E2E telecom suppliers. Transfer of knowledge, processes and routines will then enable operators to easily take over the responsibility for their network operations at the agreed point in time (if required). The proven processes, on-the-job training and certification programmes will result in high performance capability and the right competence level in operator operations organisation.

Some of the main benefits of this model are to found amongst the following:

- Shorter time to market
- State-of-the-art network performance
- Investment in proven operational practice
- In-house high performance operation capability
- Operational performance at predictable cost
- Knowledge transfer to operator staff

Hosting Model

Complexity! Control! Change! Cost! These are the some of the challenges when looking for new ways to get the most business value from telecom infrastructure investments. Scaling solutions in order to meet changing needs requires investment decisions around technology as well as personnel. In the hosting business model scenario, the supplier hosts service applications and enablers for the operator. The benefits are driven by economy of scale. The supplier delivers the capacity and functionality as desired by the operator. With this service, the customer only needs to focus on desired functions and services, while the supplier takes care of the technology, scalability, operations and maintenance.

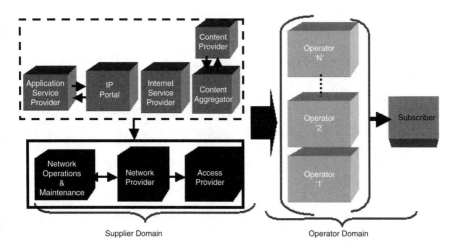

Figure 6.8 Hosting business model

The salient features of the hosting business models are to found in the following:

- Shortened time to market, from time the marketing department decided to offer a service, until it is actually is launched
- Reduction of upfront costs in order to minimise financial risks and securing a fast return on investment
- Mitigation of risk in an environment of rapid technological change
- Free up time and talent to focus on business rather than rapidly changing technology
- Ensured high quality meaning high availability, reliability and accessibility at predefined performance.

The key benefits of the hosting business models are:

- Speed
- Simplicity
- Cost efficiency

Partnership Frame Model

Access to best practices! Availability of world-class competence from day one! Guarantee of the outcome! These are the key building blocks of the partnership frame business model. In this model, the operator brings in an experienced supplier who understands all the necessary dimensions of the telecom services business in partnership mode.

The supplier comes in to join forces, providing complimentary competence to the operator in order to ensure that the operator is meeting its set targets. The partnership frame model is driven by a service level agreement where the outcome of the relationship and value are defined and agreed.

Figure 6.9 depicts the structure of such a partnership. The main aim of the model is that both operator and supplier are working towards the same targets and these are the targets of the operator.

The key benefits of the partnership frame model are as following:

- Access to best practices and processes
- Access to global benchmarks
- Availability of competence on demand
- Enables synergies between operator and supplier ensuring efficient use of people and resources
- Yearly guarantee on efficiency improvement
- Quality and time guarantee on output
- Better control over performance
- Faster time to market

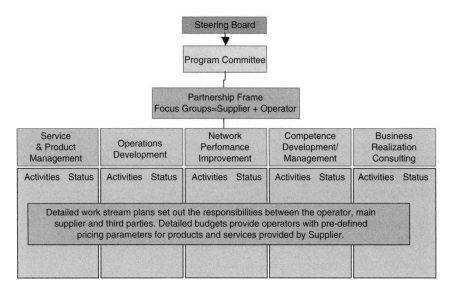

Figure 6.9 Partnership frame structure

Rural Business Model

In this example, the operator doesn't own the infrastructure, it is owned by the special purpose company (SPC). The SPC also holds a facility licence and the needed spectrum. The financial risks are spread among the equity owners while the operator keeps the market risk. In this way, the majority of the CAPEX and OPEX risks are lifted from the operator and thereby, entry hurdles for providing good quality mobile communication to the rural poor in developing countries are removed. Figure 6.10 illustrates the rural business model.

Special Purpose Company (SPC)

In the rural business model, a SPC is created for the purpose of spreading the financial, operational and market risks among the various stakeholders and through enabling a sustainable business case. The SPC functions as the centre of the business model. Being the infrastructure owner and providing a sustainable network solution for the operators, the SPC offers increased coverage and access to additional subscribers that would otherwise be difficult to cover in a profitable way. The SPC can exist in two different forms: one operator driven where the operators among other stakeholders are the equity owners in the SPC; and one neutral, with no

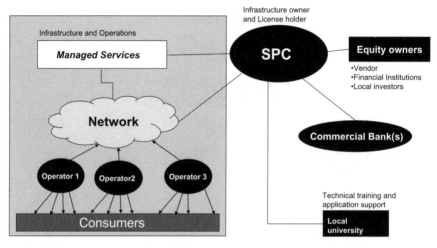

Figure 6.10 Rural business model

operators as equity owners. Which form is most suitable will depend on local circumstances. In the case of an operator driven SPC there is a risk of a conflict of interest and the main issue will be collaboration between the operators.

The neutral form, where no operators participates – at least in the initial phase – is viewed as the safest way to set up the SPC. Once the SPC is up and running, an operator participating as an equity owner could be considered.

Such a business model is adapted to local conditions and will provide the possibility to lower the MOU (minutes of use) price from today's 25–40 US cents (example from Africa) to well below 10 US cents. This is needed to stimulate traffic growth which is necessary for a profitable and sustainable mobile communication solution for the rural poor.

Technically, there are several areas that could substantially lower the total cost of ownership:

- **Site solutions**
  - Different types of antenna can lower demands on towers.
  - A number of power solutions in off-grid areas are available, but not utilised to their full potential. These involve solar cells, fuel cell technology, bio-fuel solutions and wind power.
  - New simpler masts and towers will lower site cost including civil works.
- **System solutions**
  - Local switching and local breakout could lower the cost of satellite transmission, eliminating the need for transmission with 50 to 70% depending in the region.

- Packet switching techniques will also lower the transmission cost, making satellite transmission more accessible.
- A smarter network design could lower the number of sites needed.

What if

- Satellite transmission cost could be lowered by 60–70%?
- Calls could be switched locally?
- Complete radio sites could be built 60% cheaper?
- Core network could be shared?
- Access network could be shared?
- Financial risk was shared?
- Power was a non-issue and created local jobs?
- Tariffs could be lowered way below 10 US cents/minute with maintained profitability?

Today, this is possible if solutions are packaged the right way. The name of the game is traffic growth. In poor areas, that will come with the right tariffs and the right business models.

Profitability is a key component for sustainability as discussed above, and the way to profitability in rural parts of the developing countries is tightly connected to volume. Achieving high traffic volumes among the rural poor will depend on cost/tariffs and low cost handsets. It will also depend on how much value the mobile phone can bring to the individual. That value can increases substantially above the pure use of voice if we add locally adapted SMS driven applications. Examples of such applications follow.

B2B Applications

It is in people's interest to obtain information on various topics. In developed countries, this kind of information is obtained through various media such as television, newspapers and the Internet. In developing countries where distribution and communication is limited, mobile information channels can be the answer to the problem. A user would be able to request specific information by sending a message (SMS or MMS) to an information centre. The request can be of two types, either a one-time request or a subscription request. The one-time request is replied to by one message containing the information requested. The second type of request is a subscription to an information channel and messages are received continuously. Due to illiteracy problems in developing countries, another way of accessing these information channels is via a voice portal. A user can place a call to the information central and by giving voice commands, information is read to the user. The information channels offered can be of different nature containing assorted information on specific topics.

Application Description

Consumer:

- Cleaning & hygienic information sent about personal hygiene as well as general cleaning tips
- Job notices of available jobs are sent to mobile users on demand
- Medicine information concerning a specific drug and its recommended usage
- News is delivered in accordance to the mobile user's request
- Questions & Answers: an application enabling a mobile user to ask a question within a specific subject field and get an answer
- Regional information of happenings in the vicinity of the mobile user is delivered
- Road conditions for a specific route are delivered after request from the mobile user
- Timetable information on departure and arrival times of transport is received on demand
- Weather information is delivered to the mobile user
- Yellow Pages: contact numbers can be received on request
- Commodity pricing information
- Exchange rates are received concerning requested currencies
- Industry specific information can be requested
- Market information on available amount and location of a commodity.

Key Benefits

End User Perspective

Information can be used to improve people's quality of life by giving them the opportunity to act in relation to the information given. Furthermore, job opportunities can help the individual and boost the labour market. Also, an information source such as the Yellow Pages will improve communications between people.

Social Perspective

Services bringing information to communities and their people could be used to develop societies when people become more aware and educated of the world around them.

Revenue Source

- Traffic fees for both SMS/MMS (SMS/MMS Portal) and Voice (voice portal)
- Subscription fees
- The services offered via SMS could be paid by the users either per message or a subscription fee for each information channel subscribed. Another alternative is sponsored push pull messaging applications where a third party pays for the traffic and service in exchange for advertisements in the message. The agreement with the third party could bring increased financial stability of the service depending on the business model. Furthermore, if the voice portal is used, normal voice traffic will generate even more network traffic.

Application Requirements

- Information database where all information is saved, either at one location or distributed over various information providers implementing a common interface
- Voice message recording unit for the voice portal functionality
- SMS enabled mobile equipment
- GPRS enabled mobile equipment (for MMS functionality)
- Possible employment needed for certain services.

Sustainability

Information provided needs to be in the local language or languages for this application to be successful. Furthermore, to achieve a sustainable application, the information must be of use for the people. Information providers must be trusted sources which deliver accurate information on a regular basis.

Implementation and consideration

Generally, all the mentioned information channels are requested. The differences are mainly the ease of gaining content to provide the application. Providers of news and weather exist today via TV and radio and should be accessible by contacting relevant institutions for information.

Key Steps

What steps are needed to establish a rural business model in an emerging market? In figure 6.11 we have taken the example of the so-called rural business model based on a shared access network.

As every market is unique, both the business model and the road map must be adapted to the specific local market conditions.

Lobby with Government, Regulator and Operators

Regardless who takes the initiative to build mobile communication in underserved areas and especially rural areas, it is advisable to get key stakeholders on board. If one operator has a major lead in market share it could be difficult to get that operator on board. However, normally if one operator takes the initiative to build mobile coverage in a rural region, others will come and build competing networks and we will end up with parallel infrastructures that most likely will not generate a profitable business. So, in this scenario the operators will share the revenues, but they will carry the infrastructure and operational cost themselves. The alternative would be to share both revenues and cost, thereby dramatically improving profitability and therefore, also, sustainability.

The regulator needs to support the concept of a shared infrastructure. This means that the regulatory framework and policies must be sufficient or the regulator must be willing to adopt policies to support a shared infrastructure and that the network can be owned by a third party.

In other words there is no point pursuing a shared network concept if the key players are not in favour.

Figure 6.11   Steps needed to establish a rural business model

Feasibility Study

As a first step, a work group should be put together where all key stakeholders participate. A suitable rural area should be selected and a preliminary cell planning performed to determine the number of radio sites, need for power etc.

Consumer needs must be analysed as well as their ability to pay for phones and usage. It is also important to involve local entrepreneurs and other local interest groups. For example, if electrification is needed and a biofuel solution could be used, the local farming community needs to be involved. If specific end user applications will be used, such as B2B and mobile banking, the need for local adaptation should be analysed. The feasibility study should also cover how a shared network could be financed, e.g. local versus international investors.

Business Case and Financial Statement

A business case has to be presented for each of the stakeholders in the business model. The business case must include a cash flow model and a financial statement showing the payback period. This needs to be done for the operators, suppliers and if a special purpose company (SPC) is created as the infrastructure owner, the SPC needs to be included as well.

Letter of Intent (LOI), Stakeholder Accepts

At this stage, there needs to be a written agreement that the business case is good and that the stakeholders are prepared to continue the process. From this point on, the stakeholder needs to take an active responsibility for the process. Investments may have to be done and legal papers agreed and signed.

SPC Set-up

The SPC needs to be legally set up as an operational company with shareholder agreements etc. A temporary management is needed during the initial phase to get all the internal processes in place and safeguard the ongoing set-up the business model.

Licence and Spectrum

The SPC needs to apply for a facility license. The SPC is not an operator. The facility license will enable the SPC to run radio traffic, but not to address the subscribers. Along with the facility license, a suitable frequency spectrum is needed. Obviously, both the license and spectrum issue should have been preliminary agreed with the regulator as a precondition to continuing the process.

Operator Commercial Contract

Once the licence and spectrum have been granted, it is time for the operators that should share the access network to sign the commercial contract. Also, here it is essential that the commercial contract has been preliminary agreed and that there are no surprises.

Final Planning and Roll-out

Once the commercial contract has been signed, the final network design and cell planning need to take place. The roll-out of the network will then be the last piece of the puzzle before the operational phase can start. For the operational phase, it is strongly recommended that the SPC should outsource all the operations and maintenance, instead of building up its own organisation.

Value Proposition

The value proposition of all the above business models, apart from faster time to market includes the following values:

- Improved cash flow: efficient operations packed with innovative financing
- Improved quality of network and quality of service: guaranteed quality at the predictable costs
- Cost reduction, lower OPEX: cost reduction due to efficient processes and economy of scale
- Competence transfer, increase in local telecom competence levels: transfer of operations framework and the competence to the local staff, which positively addresses the human capital of the region
- Better prices for end users: efficiently organised communications networks ensure low OPEX, which in turn allows operators to offer better prices on tariff and subsidies on handsets.

CONCLUSION

Telecom growth in developing nations is driven predominantly by mobile communication. The top of the consumer pyramid has been well served in these countries, with extremely good profits. To sustain growth and achieve similar profitability level while penetrating in the low end of the consumer pyramid telecom operators need to create a lean machine, need to now adopt new business models. Operators using traditional models to address this new wave of users will become unprofitable and their business unsustainable. The business models described here present a clear value proposition and must be tested and exploited for the good of every one in the value chain.

As operators contemplate their innovation priorities, how much attention does innovation in products/services/markets warrant? In our financial analysis, we noted a positive correlation between products/services/markets innovation and above-average operating margins. Over a five-year period, products/services/markets innovators edged past competitors' operating margins by just over 1%. Put in context, companies that are using business model innovation enjoyed significant operating margin growth, while those using products/services/markets and operational innovation have sustained their margins over time.

By adopting the business models operators will reduce the cost of service production, consumers will enjoy communications services at affordable price, and higher mobile penetration will boost the economy of the users and the state. A clear win-win for everyone.

7

Straight from Top Executives – Trends and Approaches*

OVERVIEW

This chapter makes an attempt to bring together the views of selected top executives from the major players in the communication industry and findings from recent telecommunication studies on the trends and approaches being taken to seize business opportunities being created by one of the most challenging sectors in every economy. The trends and approaches in communication systems are being shaped by consumers in both the developed and developing countries. Dynamic changes in technologies aimed at providing integrated voice, data and video services at lower unit costs with increasing speeds, quality and efficiency present both opportunities and challenges for sustainable telecommunication growth creating the need for new types of business models.

GLOBAL TRENDS IN THE ICT INDUSTRY

The key global trends that are shaping the ICT industry and are forcing the key players and stakeholders to reexamine their existing business and operations models and modify them to meet the challenges that these changes are posing.

From the past to the present, there has been a shift from disconnected communication devices to single connected devices. For about eighty

*The views and opinions expressed in this chapter are the individual author's own and should not be attributed to the World Bank, its management, its Board of Directors or the countries they represent.

Business Models for Sustainable Telecoms Growth in Developing Economies
S. Kaul, F. Ali, S. Janakiram and B. Wattenström
© 2008 S. Kaul, F. Ali, S. Janakiram and B. Wattenström

years after its discoveryy, telecommunications existed as a single large service industry offering only basic data and voice call capability. Then some 20 years ago, the mobile phone emerged as a disruptive technology. Ten years ago, the IT and communications industries were largely distinct worlds, and the Internet was new for the majority of people globally. Only five years ago did high-speed broadband communications across copper infrastructure became commercially possible. Viewed from another perspective, in the beginning there was basic mobile voice and SMS. Then, about five years ago, GPRS was introduced and with it the first possibility to offer mobile data services in a serious way (contrary to the non-serious way that WAP-based mobile data services were offered over original circuit switched GSM networks). These mobile data capabilities are further enhanced with the current rollout of 3G and HSPA networks.

This is creating a mobile data value chain, bringing content and media companies to the mobile market. Telcos used to be just big pipes, where customers provided their own content – their voice. Now we're doing other things. Some sort of service, content or proposition that is not traditionally a Telco's domain. The three blocks in the mobile value chain each have their distinct business logic.

Communication today is still a significant challenge. In a single day, you probably send and receive multiple SMSs, emails, make phone calls from your desktop and mobile telephones, and check messages in multiple mailboxes. You might participate in an audio conference call, use instant messaging and schedule meetings with your calendaring application. The irony is that rather than making it easier to reach people, the proliferation of disconnected communications devices often makes it more difficult and more time consuming. And in an age when business success increasingly depends on how quickly people can share information, this is a critical issue.

CONVERGENCE IN COMMUNICATION SYSTEMS

'In the coming years, unified communications technologies will eliminate the barriers between the communications modes – email, voice, Web conferencing and more – that we use every day. They will enable us to close the gap between the devices we use to contact people when we need information and the applications and business processes where we use that information. The impact on productivity, creativity and collaboration will be profound'. (Gates, 2006)

Convergence, long talked about, is gaining rapid momentum. The communications industry is an exciting place and this is an interesting time to be in it. Digitised voice, data and video can now be combined, changed and manipulated on a single digital platform. If the ability to merge multiple information formats on a single platform drives the desire for convergence at a device level, the availability of carrier-class IP networks, multi-service networks and software-driven switching fuel the engine for fundamental change in our industry. The transmission of voice calls over IP is a reality today and BT has no doubt that it will replace narrowband switched traffic as the preferred medium of communications in the future. We believe it poses a very real threat to established telecommunications companies like BT and others. But it also offers us new possibilities.

Mobile Telecommunications has Become an Integrated Part of Many People's Lives, Posing Opportunities and Challenges for Operators

Over the past 10 years, mobile telecommunications has evolved from a gadget for the rich and privileged to an integral part of daily life. Research by Ericsson's market research group, Consumer Lab, shows that there are many small moments during the day where people use their phone: when they get up in the morning (alarm clock, checking E-mail or SMS), on the train to work (checking e-mail, surfing webpages, playing games), at work or school. It is very much about understanding this timing of activities, locations and contexts during the day. As one interviewed CEO said:

> 'The fundamental building block is that the customer has got a habit – a habit of car-rying the device with him. In our country, the device is switched on most of the time. It's even switched on while charging at night. It is literally with me, by my side, 24 hours a day. We have been fortunate as an industry to build up a customer habit in the last 10–15 years, to the point that, when people leave the house without a phone, they feel that something is missing.'

Another interesting observation from the Consumer Lab research is that mobile usage is very much about staying in touch. Not only to make announcements, but increasingly to share experiences with your closest family and friends. Rephrasing Descartes: 'I'm connected, therefore: I am'.

OPPORTUNITIES AND CHALLENGES PRESENTED BY THE MOBILE PHONE

Opportunities:

Data on Consumers

One opportunity is related to the fact that the phone is personal and always switched on. This gives the mobile operators a treasure of information about the user that can be used for marketing and Customer Relationship Management (CRM) purposes. As one executive explains:

'We have built this habit and network intelligence to have the ability to know what you're doing. I know when you leave the country, where you are, when you go to bed -when you're likely to be stationary at one spot for 6 hours- when you are likely to be more active, who you make your calls to, which country you make calls to, when you make your calls, what you're browsing and what kind of service you want, what kind of offers you respond to. As an industry, I don't think there's any parallel in the amount of data we have on our customer. Then the question is, with that relationship, that habit and with billing – the ability to charge money – what can we do with it?'

The question in this case is: to what extent will privacy laws allow operators to use subscriber behavior data?

Contact and Messaging

Another opportunity is to leverage on the focus on contact and messaging when developing new services. A clear signal from market research is that it is not *content* that is King, but *contact*. IMS enabled services where voice calls and sharing pictures or clips are combined, tap into this preference. A third opportunity, especially for pure mobile players, is increasing voice usage at the expense of fixed voice services. Experience from some advanced markets shows that there is a large pent up demand for cheap mobile voice. As one CEO describes:

'The younger market is very relevant. They are unquestionably the early adopters. Anyone in their 20s who is flat sharing and relies more on their mobile as their one and only point of contact belongs clearly to the leaders in terms of fixed to mobile substitution. The trend over the last few years has been dramatic – as dramatic as anything you have seen in the mobile marketplace. I think that trend around targeting fixed line minutes will become more pronounced over the next few years because there is a significant opportunity for a market that gets increasingly competitive.'

Challenges:

Maintaining Service Quality

On the challenges, operators have a large responsibility not to disappoint their customers when it comes to service quality and availability. Customers will take it personally when their phone is not working! Another challenge is related to mobile data services. As one CEO describes:

'Our market has been and still is very voice-centric. Which means that the decision-making process for the average consumer is based on which handset they like? Secondly, they check with the shop on which operator offers the highest amount of subsidy for the phone. Thirdly, when they have the subscription and the phone, they then ask what they can do beyond voice calls.'

Changing this behavior requires a large effort.

Consumer Behavior

Even Greenfield 3G operators take a voice centric approach for the moment with the possibility to up-sell data services. *'To change people's usage patterns on voice is easier than changing people usage patterns on a new range of services. That is certainly the way we have pushed the market place, rightly or wrongly, and that's the way we have seen things going. The market for data services is not that far away, and maybe in the next few years we will be in the right place, but there is enough penetration of content and other multimedia services to enable you to almost lead with those aspects of your service in terms of acquiring customers but I don't think that is the case today.'*

Challenge of Meeting High Expectations

Another challenge is that, as customers expect things to work for them, services have to be simple and reliable in all aspects. This has to do with the interface and navigation on the phone, information and support provided at the retail level and getting all the details right to deliver and bill the service. The challenge in understanding mobile users is nicely worded by one CEO talking about how his market thinking evolved over time:

> *'Our approach was customer-centric. So, we need to understand our customers and provide solutions to engage them. We then moved from "customer-centric" to the "demand approach". This means that we need to understand the customer more than they understand themselves. We asked many things from our customers and found out that they don't know. Because we are offering them services, which they have not seen in the past, it is hard for them to imagine. This means that we should educate the customer and not let the customer educate us. But we still have the customer at the center of our attention.'*

Understanding end user behavior and its drivers is the beginning and end of all mobile business and probably one of the toughest tasks around.

APPROACHES BEING USED BY MAJOR PLAYERS IN THE COMMUNICATIONS INDUSTRY

Microsoft's Person-Centric Communications

The arrival of unified communications signals the beginning of the convergence of VoIP telephony (which provides the ability to route telephone calls through the Internet), email, instant messaging, mobile communications and audio and video Web conferencing into a single platform that shares a common directory and common developer tools. Unified communications also takes advantage of standard communication protocols such as SIP (Session Initiation Protocol) to route communications to the right people on the right device.

Building on these communications standards, Microsoft is delivering a powerful set of unified communications capabilities that provide the framework for person-centric communications across locations and devices. The result is an approach to unified communications that is:

Personal and Intuitive

One of our most important goals is to make communication and information access seamless and personal, no matter where you are or what device you are using. Presence – which provides information about your availability – will enable you to reach the right person on the first try. Intelligent information agent software that understands how you prefer to work will give you control over who can contact you, on what device and at what times. SIP standards and software-based call management will make communications richer and more intuitive, and provide seamless transitions from one communications mode to the next.

Convenient and Integrated

Today, when you contact a colleague, you probably need to switch from the application you are working in to an address book and then to a device (like a telephone) or a different application (such as e-mail). Microsoft unified communications will enable you to collaborate directly from the application where you are working. Integration with Microsoft Office will help make Microsoft Outlook the center for all types of communications experiences and provide seamless access to collaboration tools such as Microsoft SharePoint. By delivering a standards-based platform, Microsoft will enable developers to integrate communications into applications that provide even greater value, convenience and power.[1]

Flexible and Trustworthy

'Microsoft unified communications will enable organisations to consolidate their communications systems into an integrated platform that utilises a single identity for each user and provides a common management and compliance infrastructure. This will enable IT departments to significantly improve communications and collaboration capabilities

---

[1]The Unified Communications Revolution. See http://www.microsoft.com/mscorp/execmail/2006/06–26unifiedcomm.mspx last accessed February 2008.

while reducing complexity and lowering total cost of ownership. Built on a platform that is secure and reliable, Microsoft unified communications technologies are already helping leading companies achieve groundbreaking TCO. Ebay, for example, has lowered its per-mailbox costs by 70 percent. At Nissan, collaboration technologies have helped save more than US$135 million. And Siemens has unified 130 business units into a single Active Directory' (Gates, 2006).

British Telecom's Radical 21st Century Network Approach

BT considered multiple strategic options before deciding to follow an innovative approach. BT arrived at this decision because they foresaw multiple growth opportunities in new wave businesses, in information and communications technology and in mobility and broadband. BT had a high need to invest with the intent of capitalizing on the new wave opportunities. This has resulted in BT's strategic response that has culminated in the 21st Century Network (21CN).
   21CN is

- a multiservice end-to-end IP network with an integrated systems stack to support it, which would deploy common capabilities as the basis for future product development;
- a converged world where customers' experiences are simple and complete, where customers have direct control over the way they choose, use and change the services they use when they want. These services will improve customers' quality of life and help make their professional and personal lives easier and more productive. Simplicity is the key;
- a broadband and multimedia world where a customer can access any communications service from any device from anywhere – at home, in the office, on the move – all at increasing broadband speed. It will offer a rich array of information and entertainment services to whatever device the customer chooses. A handset will be a fixed line phone when you're at home or in the office and a mobile when you're somewhere else.[2]
- providing other operators with access to its functionality and intelligence, as an IP domain, enable anyone to build applications and in the IT-centric, IP communications world, it will be cheap to experiment;

[2] Project Bluephone which is in trials (expected to be launched in 2007) is a converged voice/data communications service offering transparent fixed/mobile service differentiated only by available bandwidth.

- the most complete, exciting and ambitious business transformation underway anywhere in the telecommunications world today. It means BT will be the first incumbent operator to switch off the PSTN and go all IP from end to end;

Delivering the 21ˢᵗ Century Network Required Collaboration Between Major Players

BT today (25 April, 2005)[3] took a major step towards delivering its 21st Century Network with the selection of its preferred suppliers. The 21st Century Network is the world's most radical next generation network transformation programme. It will require an investment by BT of up to £10 billion over the next five years and will put the UK at the cutting edge of innovation, providing industry with a real competitive advantage, and consumers with a world-class communication service for the 21st Century. This was a culmination of two years of discussions and negotiations with over 300 potential technology suppliers from all corners of the world. This is has been an extremely competitive process in what is one of the largest single procurement programmes ever undertaken in the communications industry.

A final list of eight preferred suppliers has been chosen to work with BT in five strategic domains:

- Fujitsu and Huawei have been chosen in the access domain which will link BT's existing access network with the new 21CN
- Alcatel, Cisco and Siemens have been selected as preferred suppliers for metro nodes which provide routing and signaling for 21CN's voice, data and video services
- Cisco and Lucent will be 21CN's preferred suppliers for core nodes providing high capacity and cost efficient connections between metro nodes
- Ericsson was selected in the i-node domain – in essence the intelligence that controls the services
- Ciena and Huawei have been chosen in the transmission domain to supply the optical electronics that will convert the signals carried at high capacity over the cables connecting the metro and core nodes

The selection of these eight suppliers will allow dozens of smaller and innovative subcontractors to become involved in the delivery of the 21st Century Network. The programme will attract significant new investment and employment to the UK.

Matt Bross, BT Group's chief technology officer said: *'The capability that BT is putting in place through this investment in 21CN is unequalled anywhere*

[3] http://www.btgtm.com/BTGlobalWholesale/Downloads/21CN_suppliers.doc

in the world. It will enable us to introduce new services at a speed that is simply impossible today.'

Paul Reynolds, BT Wholesale chief executive, the BT Board member responsible for the programme, added: *'21 CN is a key infrastructure that will fuel the UK economy and provide a flexible way for consumers to use new services. The selection of the preferred suppliers is an incredibly important building block towards that vision. 21CN will also radically reduce BT's cost base, with identified savings of around one billion pounds a year.'*

Cisco's Approach – Delivery of New World Telecommunication Services Which is Internet Based Rather than Circuit Switched Networks

As the Internet takes hold in the telecommunications arena, according to John Chambers, CEO of CISCO systems,[4] it will enable telecommunications companies to deliver a mix of New World data, voice and video services, helping all companies run their business more effectively, sell their products and services on-line, and communicate better to customers, vendors and employees. In addition, the adoption of New World technologies will help level the playing field for all nations. Voice is no longer the single driver to conduct business today, Data will soon exceed voice traffic over many international circuits, and the Internet has emerged as the new driving force. New World telecommunications services are delivered by new network infrastructure that is Internet-based rather than circuit-switched. These new IP-based, multiservice networks can carry data, voice, and video on a single network, reducing costs and extending reach of businesses to worldwide markets. Multiservice networks not only provide a means to carry voice traffic far more cost-effectively than traditional circuit-switched networks, but more importantly, they pave the way for a variety of lucrative value-added services. Such services are becoming increasingly important to telecommunication service providers, especially those operating in markets where aggressive competitors are looking to capture the profitable long-distance, leased line and value-added services. These services – which range from managed services to Virtual Private Networks (VPNs) to total system integration for corporate customers – offer a valuable means for service providers to move up the value-chain. Most experts now share Cisco's view that the integration of data, voice, and video traffic will be based on Internet-based networks, not circuit-switched voice networks.

[4]Chambers, John. 1998. Speech in Beijing, China. CISCO systems. September 28, 1998. Link: http://newsroom.cisco.com/dlls/fspnisapic0c9.html

Telefónica Chile's Triple Play Strategy of Video, Voice and Data Over One Broadband Platform

Chile is the regional pioneer in terms of convergence, with VTR Global-com having been the first operator in Latin America to develop the triple play strategy, which combines video, voice, and data over one broadband platform. Fixed-line incumbent Telefónica Chile entered the triple play market in June 2006.[5] Chile's Internet and broadband penetration rates are the highest in Latin America (apart from a few Caribbean islands). Several operators provide broadband access using ADSL, cable modem, and wireless.

External Collaboration is Indispensable

According to the *IBM Global CEO Study* 2006, based on a sample of 765 Global CEOs of which there were 43 Telecom CEOs, concur with their fellow CEOs in other industries that external collaboration is indispensable.[6] Eighty percent of telecom respondents acknowledge the importance of collaborating with a wide range of external partners, especially as converging services and industries reshape the business scene. Indeed, they say that 51% of the ideas they develop come from external sources.

But Are They as Collaborative as They Need to Be?

Survey findings point in the other direction. There is a marked gap between the significance telecom CEOs attach to collaborating with others and the extent to which they succeed in doing so. Only 50% say that they are strong collaborators. They cite complex processes and inflexible infrastructures as two of the biggest obstacles their companies face in integrating new technologies and collaborating with external organisations. However, if telecom providers are to deliver the full 'quad play' of next-generation services over IP, they will need to manage a much larger portfolio of services than ever before and be more responsive to a broader array of customer requirements. This will inevitably require telecom providers to work with a much wider range of partners and suppliers, across the value chain. It will also require them to adopt a more collaborative approach to the integration of new technologies and embrace open standards.

---

[5] Information obtained from link: http://www.budde.com.au/Reports/Contents/ Chile-Key-Statistics-Telecom-Market-and-Regulatory-Overviews-2367.html
[6] Source: *The IBM Global CEO Study 2006*. IBM

Declining Prices of Telecommunication Services

Prices of air tickets and telephone minutes have come down dramatically, sometimes to 5–10% of their levels 15 years ago – as a result of liberalisation and competition. What happened to the fixed telecom market seems to be happening to the mobile side of business as well. Strong price pressure on basic mobile services like voice, but also SMS, is a trend that is also quite evident in Asia and increasing.

Prices are driven down by a combination of factors that vary from country to country. Among the reasons are:

- The entrance of new players into the market. Having to build up a customer base from scratch, the favorite tool-du-jour for these newcomers is price, especially in markets where prices of mobile services are relatively high. As one executive put it: *'Another way of looking at it is the EBITDA margins of the incumbent operators. They are still very high. A new operator will say; I am happy with half that EBITDA. So how do I achieve that? I'll just halve the price.'* Price pressure from a new entrant is even higher if that entrant comes in with a technology that is superior to what is available in the market. In the case of new WCDMA networks, capacity is much higher and unit price much lower than of GSM networks. It makes sense for a new WCDMA entrant to lower prices to an extent that people start using the mobile phone instead of the fixed line services. The combination of low prices and high usage puts GSM operators in a tight spot as they have to choose between expanding capacity in their GSM network or rolling out WCDMA networks themselves. The impact on existing GSM markets can be quite large: *'The dynamics of another player in the market place – an aggressive 3G player turned that whole slow steady GSM maintenance position on its head. What has happened in the last 12 months as the operator started to gain market share has been a significant alteration in people's plans for 3G.'*
- Many markets are reaching a level of saturation. In developed markets 80%–90% of the population subscribes to a mobile communication service. As one executive puts it: *'The market is slowing. I'm not talking in terms of trying to announce nice customer numbers going north. I'm talking in terms of absolute dollars you get from customers in the mobile market place'.* Even in less developed markets, mobile penetration will stabilise in the next few years. Official penetration figures can be deceptive due to double counting. In some countries actual take-up of mobile services could be 10%–20% lower than official figures due to ownership of multiple SIM cards and/or double count of subscribers churning between operators. With market growth gone as a driver for top-line growth operators find themselves in a

competitive situation where the temptation of dropping prices is irresistible.

Approaches Taken:

Price Wars in Limited Marketplace

In a marketplace with two or three players it makes sense from an economic point of view to maximise revenues by not engaging in any form of price war; in other words, nobody wants to spoil the party. Although mobile markets fit this description, operators sometimes do start price wars. One description of this irrational competitive behavior from a market with two established and one new player: *'They don't make rational market moves that are value enhancing from an industry perspective. It is literally; 'you hit me, I hit you' price wars, marketing wars, trade wars and so forth.'* This comprises a fourth driver for declining voice and SMS tariffs.

Seek Alternative Growth Opportunities

The obvious reaction of mobile operators to falling voice and SMS revenues is to protect these revenues as much as possible by seeking alternative growth opportunities with existing customers and to start looking for business in other markets. One way to protect existing revenues is to differentiate from the other guy. One differentiation strategy mentioned by multiple executives is to build a strong brand. As one CEO put it: *'So what we're looking at now is to become an identity in the eyes of the customer – the brand value of our company. What we want to stand for is that we are a one-stop shop for everything to do with communications.'* Another way to differentiate from the price fighters in the market is to create a superior customer experience. To do this, the operator must have each step in the delivery of the service right.

Work Closely and Understand the Consumer

As one executive put it: *'Have we thought about the fundamental issues of devices having to work properly, the kind of phone settings required? That's why we're trying to come out with a sound proposition to the customer, that's what we're teaching customers how to use the phone. It all comes down into execution details and unfortunately it gets down to a very low level in order to make it work.'* This is something that the industry is traditionally not strong at given the past experience with mobile data services.

Enhance Marketing Functions

Protecting existing business against downward pressure on service prices requires improving and enhancing the marketing function by operators. Better market segmentation and a continuous redefinition of the customer offerings can increase total revenues. Executives say they have to *'get closer to the customer'* or even *'Winning the heart and get their emotional attachment'*. In the words of one CEO: *'I think the learning curve that the market place is going to go through around how you develop, deliver and package, promote those services to consumers is going to be a huge challenge.'*

Acquiring Own Retail Channel

Some of the operators interviewed see it as crucial to take control of the customer interface, often by building or acquiring their own retail channel. One executive that actually acquired a retail chain to this purpose comments: *'To prepare for this new environment, we changed our strategy for distribution. We pulled out the distributors and moved everything in-house. I don't believe that you can rely on the distributors to do it for you, especially when you go into the world beyond voice. You need great displays, rule of engagements with the customer, demonstrating the service, educating the customer and setting it up. It's one customer at a time at the frontline.'*

Trial and Error Method for Developing Consumer Oriented Services

With voice revenues dropping it is natural for mobile operators to look at data services to fill the gap. In the words of one operators executive: *'So, we will continue to introduce some of these services that we think will have a positive impact on revenue and those we think are easily scalable to all our customers. This is just one example because we have many more in the pipeline that we believe will have a positive ARPU and revenue contribution.'* Although talk about finding the holy grail of the *'killer application'* has not diminished over the years, the executives who were interviewed had a more pragmatic view on the subject of finding the right data-service portfolio: *'Instead of trying to do intensive analysis before implementing anything, we implement first and see what the customers like. If they like a service, we expand it. If they don't we stop. I believe in experimenting. So, from the touch points, we get our inputs. These inputs give us our framework or guides, not the final answers. We launch 10, they may like 1 and we stop 9. And then, we launch again.'* (See example of Wireless Application Protocol – Box below.)

Example of WAP

Remember Wireless Application Protocol, WAP for short? Most people will associate WAP with overselling and under-delivering on the mobile internet promise in the late 90s. What most people do not realize is that WAP is still there and actually quite successful. Part of the comeback story of WAP is that access and usage of WAP based services has been improved substantially in terms of e.g. accessibility (one-button access), navigations (improved menu structures), pre-configuration and appeal (colors).

Simplicity of mobile data services is seen by most executives as a prerequisite to develop the demand for mobile data services in the coming years. *'The second important thing for us is that it has to be easy to navigate. When we started 2 years ago, I used to say to my guys that it takes a PhD to get it to operate. So, we made sure that our services were easy to navigate. Pre-activations and activations-over-the-air for the handsets today make it much easier for people to grasp and to try out.'* Controlling the way information is presented and organised is an important part of this strategy and operator defined handsets the way to do it. The operators increasingly want to set requirements on things such as menu structures, preconfigurations, features and branding of the phones:

> *'Things are starting to change. We have more handsets being tailored to the operator's requirements to make the use of mobile data easier. Something I have been doing for the past 14 months is to lower the barriers to usage for the customers. First thing I did was to fix the handset problem. I don't have a better word to use it – just simple. If we can make it so simple, that the customer does not need to learn and can use it so effortlessly; we have reached our objective.'*

This creates an interesting tension between operators looking for smaller badges of customised phones and the handset vendors looking for economies of scale. Large global and regional operators can use their size to put pressure on phone vendors. Smaller operators have to work with local phone vendors, such as in Taiwan, or form alliances to increase their purchasing power and set requirements.

Redefining the Operating Model

It might take a redefinition by the operators of their operating model to open up new ways of doing business. One alternative way of looking at the mobile business is to see operators as investing in network assets and capabilities that can be packaged in many different ways to address as many different needs in the market. As long as the incremental cost for creating and delivering the service is lower than the incremental revenue that can be charged for the service it creates value for the company. The enterprise market is seen by some executives as the most obvious place to look for such opportunities. In some cases the mobile channel provides a cost efficient and effective alternative to steps in other industries supply chains. One example that is happening today is the packaging, distribution and charging of music over mobile networks. Another example is the success of mobile transactions using the mobile phone, where retailers see the mobile channel as a cost efficient alternative to cash without the threat of the other party taking over their customers. In some countries, the lack of a proper basic fixed telecommunication structure provides opportunities as one executive describes: 'If you look at the basic infrastructure for data for the fixed-line – be it fiber, DSL, leased circuit, ATM – it is extremely under-developed, extremely challenging. This leaves a service gap in the industry. Looking into the future, 3–5 years down, then the mobile data service becomes a viable alternative to the fixed data service. That's when you talk about 3G coming in to become a real alternative.' WCDMA combined with HSPDA was mentioned several times as a potential alternative to ADSL for broadband fixed/portable internet access.

SEEKING OPPORTUNITIES TO INNOVATE

Officially 8.2% of the GDP of the Philippines comes from money sent back home by nationals living and working overseas, mainly in places like Hong Kong, Singapore and the Middle East. Taking into account the money that is repatriated through unofficial channels such as Western-Union, money courier, post office type of remittance, the figure is even higher. The same situation, with different figures, applies for countries like India and Indonesia. These overseas communities provide another opportunity for mobile operators. One executive describes it as: *'To go on a very targeted basis offering overseas nationals type of mobile schemes, special pricing packages for our country, the ability to transfer or buy prepaid load in Singapore send it to your relatives here'.*

Operators are Finding Different Ways to Drive Down Cost, Redefining Industry Relationships in the Process

As with every industry that is reaching maturity and facing saturating markets, the focus of mobile operators moves from taking market share and acquiring new customers to controlling cost and retaining existing customers. Reducing capital as well as operational expenditures is a prominent part of the strategies of all interviewed executives. And it is not only about reducing cost; sharing risk with vendors, content providers and even competitors is an equally important part of these strategies. Capital expenditures involves the operator's relation with its vendors. Operators do not share the same opinion about their relation with their vendors going forward. On the one hand some operators see ever closer co-operation with fewer vendors that have complementary portfolios:

> *'I think the operators will have fewer, more intense partnerships with vendors in the marketplace, because they see the complexity of integration in the network. They see value in aligning around perhaps what I call their own 'ecosystem' of suppliers. We want to make sure that we have the right relationships with the companies who have been with you through the journey and help you transition from one technology platform to another.'*

On the other hand some operators take the opposite approach, taking vendors on a contract-by-contract commodity basis:

> *'Many contracts in the past were based on a turn key base. Now we literally go to Ericsson and say you only give us equipment prices, we will directly talk to the suppliers of towers, the suppliers of transformers and generators, backhauls, civil works, steel manufacturers etc. and take out each one on a commodity basis. And I will project manage the build.'*

Another strategy to reduce investments is to share parts of the infrastructure.

Although, these 'shared networks' have only been implemented in Australia, the concept is considered by other operators in the region as well. In the words of one CEO with experience on the subject:

> *'The sharing of costs and reduction of costs that I talked about in terms of a maturing marketplace will still be very relevant going forward. So the infrastructure sharing agreements you have seen in this marketplace are just a huge flag waving about the changing of behavior between competitors. There is more of a realisation that no one*

is going away; everyone's got to make money. So let's work with each other in areas that are non-differentiating to try and reduce cost.'

Saturation of the mobile market and with it the change in competitive behavior, away from buying market share, creates new opportunities for network sharing. As one executive describes: *'We will even now look at issues related to network sharing and so forth, while in the past rivalry is so intense that this was not an option. But at least conceptually these are things we are prepared to explore.'*

Shared networks are one way for mobile operators to create economies of scale for their (network) operations, creating savings in both capital and operational expenditures. Another option to reach the same effect is outsourcing of network operations. The logic here is that e.g. a telecoms vendor with multiple customers in one area can create economies of scale by running multiple networks that individual operators could not achieve by their own. In order of complexity, activities that are outsourced in this way are: roll-out and site acquisition, field maintenance, network operations, network design and optimisation. Here again, different operators take quite different positions when it comes to outsourcing, ranging from one operator that has actually outsourced their operations, to moderately positive:

'I think if you are a business leader running a network group or a technology group like myself, you would see this as an option available to you. It could lead to full end-to-end life cycle and operations outsourcing of your network elements. If you are an engineer or a long standing Telco person you are probably threatened by it.'

to a more cautious outlook that is linked to a changing mobile value chain:

'Outsourcing – not a chance with us, not as we speak. But I can see where you're coming from. If we move over to another dimension and talk about the convergence of both technology and services'.

Full control of the service delivery to the customer is cited as the most important argument against outsourcing. As one CEO explained: *'I want some level of control in how I engineer my network. I want control over coverage and responsiveness, and we're weighing on that all the time.'* Even if SLAs are defined and KPIs are met, there is a difference in commitment between own staff and a third party; not measurable, but perceivable by the customer. One CEO that outsourced their customer call center and later took it back in stated: *'They weren't keeping the same quality, they weren't flexible and the costs were roughly the same, so I said this is just kidding me.'*

The third way to create economies of scale, after network sharing and outsourcing, is to create mobile alliances. In APAC there are three such alliances: Asian Mobile Initiative (AMI), Bridge Mobile Alliance and the I-mode alliance. Operators from all three alliances participated in the interviews. Creating buying power versus content providers and handset manufacturers, especially for smaller operators, was cited as the two main cost benefits of membership.

It's the Way That You Do it . . .

As typically 30% of total operating expenditure (OPEX) comes from operations (non-depreciation), it makes sense to look at people and processes to bring down cost. The drivers for this process re-engineering vary, depending on the operator. For one operator in a strong growth market, the challenge is to increase output in terms of SIMs per month and keep that output stable:

> 'We used to be able to push out only 200 000 SIM cards a month in the market in 2003. Everything from distribution to logistics was re-engineered. We ramped up our SIM card distribution in a few months to 500 000, then to 1 000 000 and we're currently at 2 000 000 a month. How did we do that? We couldn't just be putting more people. We had to change the rules. So, we changed the rules. We worked with the suppliers and we worked with our distributors. So now, we're consistently pushing our supply out without fail.'

For another, incumbent, operator the challenge is to create consistency in operations over multiple (fixed and mobile) networks:

> 'There are probably 6 to 8 key processes within each telco. Each telco traditionally has been looking at itself in the sense of vertical silos. Here is my mobile group, here is my data group, here is my broadband group, here is my fixed line voice group and we treat them separately. We are trying to view the company not from a product or network silo perspective but view the company from an end-to-end process perspective. So there is a customer facing process, there is a billing process, there is an activation and assurance process, there is product development process, there is a technology delivery process, and there is a procurement process, ie. looking at it horizontally. We then say, if we have a common approach to those processes no matter what organisational structure we have, no matter what is the nature of the services, then we should be able to take cost out of the business and be more efficient in providing services to our customers.'

THE THINGS WE ALL WANT TO SHARE

Risk is cost, as all businessmen know, and it is natural in a maturing industry that risk is redistributed to those players that can absorb that risk best. From the interviews, three types of risk sharing occur: between operators, between operators and their vendors and between operators and content providers. To start with the last; to an operator it does not make sense to pay the full premium for content upfront:

> *'The idea that you write a big cheque to a content provider to get rights to a piece of content doesn't make sense to me when its not proven that customers will use that content. So a formula that is more of a revenue share is more appropriate, or has fixed payments according to the number of customers. Those types of models are more relevant going forward.'*

This is especially true if it is the packaging and presentation of the content that makes the difference to the end user, and not the content itself.

Almost all operators mentioned risk sharing with vendors. Creativity is required from the vendor, as one CEO stated: *'It means the vendors themselves have to be prepared to engage the operators on a very different level, looking at even creative ways of pricing, licensing, costing, design, specs etc. in order to bring down the cost structure as much as possible.'* Or as another CEO more bluntly remarked:

> *'I think the models around partnering are going to change because as the business model gets tougher and tighter, all the operators are going to look to suppliers for some skin in the game and that is a hard one for a lot of people to get their heads around. So the idea that you are going to toss over a whole lot of boxes and we're going to take all the risk and we're going to pay reasonably high prices to do that is nonsensical in a competitive marketplace and a new wave of technology.'*

However, operators also realise that risk has a price and that there are limits to how much can be pushed to others. I always believe that nothing comes for free. It is like insurance, the more you put the risk on the insurance company the higher your premium.

'Philosophically, as an operator we have to be prepared to accept that. Somewhere there is a balloon that if its costs are squeezed here it has to come out somewhere.' In the end, it is economic forces at work, driven by increasing competition in a mature, reasonably transparent market. Forces that drive a rationalisation of the mobile value chain, where economies of scale and propensity to absorb risk determine who does what with whom and for what price. No escape from that!

To protect margins in the face of falling prices for basic services, operators have to take a hard look at their cost base. Reducing capital and operational expenditures requires re-definition of the relation with vendors; either by having ever closer co-operation with fewer vendors, or by taking each vendor out as a commodity, driving down unit cost and project manage in house. Another way to bring down cost is to create economies of scale e.g. by sharing non-differentiating parts of the network with competitors, outsourcing operations and maintenance to e.g. vendors or by the creation of mobile alliances. Apart from bringing down expenditures, reducing risk is equally important. Risk sharing agreements with vendors, content providers and competitors is increasingly seen as a standard practice.

Changing Telecom Revenue Generating Business Models Driven by Insatiable Consumer Demands for Multiple Uses of Single Communication Device

Past to the present – Changing revenue streams – from simple data services to complex voice, video, music, games, messaging, network services – and changing technologies – increasing adoption of broadband and wireless technologies – Verizon, USA.

Five years ago, Verizon had 1.6M broadband customers. Less than 10 percent of telecom revenues came from data.

Today, Verizon has over 7M broadband customers, hundreds of thousands of video subscribers, and – for the first time in a long time – consumer revenues are growing.

Five years ago, Verizon had 32.5M wireless customers, who averaged just under 400 minutes of use every month. Wireless was almost entirely a voice service, with data accounting for a little more than one percent of revenues.

Today, Verizon almost doubled the number of customers, to 60.7M, and grown minutes-of-use, to more than 700 a month. But look at these statistics from the first quarter about the fastest growing services – 22.3 B text, messages, 450M picture messages, 106M downloads of music, videos, games and ring tones. None of which existed a few years ago.

Five years ago in the enterprise space, data and Internet made up about 25 percent of Verizon's revenues.

Today, it's 50 percent. IP and managed services are growing at a 25 percent clip. Verizon's fastest-growing service is Private IP, which was just launched five years ago. And fully 40 percent of Verizon's customers opt for the company to manage their networks for them – by hosting their servers, managing their security, even managing their IT applications. All of these new services have boosted Verizon's revenue growth from a flat rate five years ago to four percent today.

Growth of Broadband Subscribers – Nippon Telephone and Telegraph (NTT), Japan

The numbers of broadband and mobile phone subscribers in Japan have been changing over the years, as shown in Fig. 7.1. The total number of

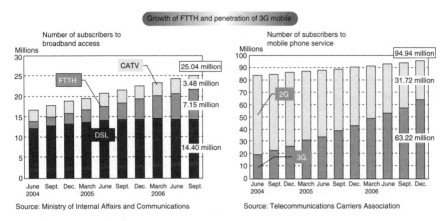

Figure 7.1   Changes in the numbers of subscribers to broadband access and mobile phone services

broadband subscribers has continued to grow, and the 25-million mark was passed in September 2006. As indicated by the breakdown by access method, the number of DSL (digital subscriber line) users began to decline in June 2006, while the number of FTTH (fiber to the home) subscribers has been growing rapidly signaling the advent of the age of optical broadband access. (2007)

NEW 3G NETWORKS POSE BOTH CHALLENGES AND OPPORTUNITIES FOR MOBILE OPERATORS

There are some distinct differences between the introductions of GSM back in the 90s and 3G now that make the latter, if nothing else, more complex. First of all the market into which 3G is born is quite different from the market that GMS was launched into. Back in the early 90s it was the beginning; everybody started more or less with a clean slate. Now all operators, save the 3G green fielders, have a legacy network and a subscriber base to consider. Something the green fielders are well aware of:

> 'You have big incremental costs and marginal incremental revenue. You've still got to deliver the operational capability. You roll it (3G) out because someone else is doing it, their doing it well and they are pushing the whole market place. By doing it, you're doing it from a defensive standpoint, not an aggressive acquisition standpoint. You've got your base already. Your challenge now, having rolled out 3G, is how do I manage the rate of migration of my customers from here to there. When customers migrate, quite simply, it means cost and marginal incremental revenue. The costs are around the subsidisation of the customer moving over but the other cost is that when customers start looking around, they don't just look at you.'

Research by Ericsson Business Consulting, together with an Asian operator, shows that if an operator decides to only issue 3G phones from a certain point in time onwards, taking into account natural churn patterns, that operator can migrate its full subscriber base from GSM to WCDMA in 4 years time and close down the GSM network. This providing that the choice in model and price of WCDMA handsets matches that of GSM. If the operator wants to migrate faster it will cost big money in marketing and subsidies. The upside of an existing subscriber base is that it provides an operational cash-flow that can (partly) finance 3G investments, reducing the financial risk.

When GSM came to the market there was no alternative available. Today with the launch of 3G there is a wide variety of standards and technologies aspiring the same spot in the sun. WiFi, WiMax, OFDM and TD-CDMA promise more or less similar or better performance at a lower price then WCDMA. Although none of the interviewed executives believes these technologies to have a superior value proposition, they do see potential disruption in the market, affecting the value of their 3G investments. One CEO puts it as follows:

> 'They can be disruptive for real, but that is something that awaits discovery. But they could be disruptive because their promises can attract dumb capital and do a lot of damage – distort pricing, confuse customers – even though it finally fails. Some of these things written about seem alluring, though it's not all there. But you're looking at the future and it could be. But if someone is willing to put $100 million and let you have a try, it can have an impact. And so we need to think carefully about it.'

Lower capital expenditures for initial network rollout of these technologies is only part of the story, and probably the easy part. The challenge for new players coming to the market with alternative access technologies is to build a brand and customer base in a saturated market and set up operations for their networks. The same CEO again:

> 'The equipment is cheap, but the administration, that's what kills you. The salaries you have to pay, the rent you have to pay every month is what kills you. Not the equipment price. The backend of doing business, like my costs in doing marketing, customer service, sales, having retail shops. Spending money for the equipment is the easy part.'

Apart from market conditions, legacy issues and alternative technologies, a fourth difference with the introduction of GSM has to do with the type of services offered. 3G is tightly linked with mobile data services and with it the creation of a mobile data value chain. Bringing content providers and service providers into the play complicates the business models for mobile operators compared to the GSM voice only models of the old days.

So who is better off? The early birds or the laggards? The early introduction of 3G to the market limits legacy problems and helps the operators to differentiate, positioning themselves in the market. Later introduction means cheaper and more choice of handsets and the opportunity to learn from others. A clear case of a half glass of water?

Opportunities and Challenges

Operators see the introduction of 3G as both an opportunity and a threat. The technology brings lower unit cost and higher capacity, enabling operators to differentiate themselves and develop new businesses. At the same time 3G requires large investments when the market might not be ready for the type of new services offered. 3G can be positioned as; the enabler of a fixed mobile substitution strategy (cheap voice), the platform for new, exiting premium mobile data services for consumers and enterprises and as an alternative for fixed broadband access in countries with poor fixed telecommunication infrastructure. Compared to the introduction of GSM in the early 90s, 3G faces bigger challenges due to different market conditions, legacy networks and subscriber bases, alternative technologies and more complex business models.

It seems that, as spectrum for next generation mobile services is licensed and networks are rolled out, 3G receives contradicting reviews from operators. The ones that see the glass as half full argue that 3G brings the opportunity for differentiation from the competition. For these operators 3G brings more speed, more capacity at a lower price, enabling them to sell cheap voice and exciting new mobile (data) services. For those who believe in the glass being half empty, it is a new technology, pushed by vendors and regulators, that they are forced to invest in. Creating a return on that investment is their big challenge in the coming years. In the words of a one CEO: *'All the big 3 operators have a pragmatic view in rolling out 3G. Our market is still predominantly a voice-centric market. We have to launch it and we agree that 3G is the future, but in the short-term, it's not going to be an exciting venture in terms of revenue because we still cannot find the additional revenue to cover the extra costs.'* Asked where 3G revenues should come from, operators give three pointers: voice as part of a fixed mobile substitution strategy, mobile data services like today but faster, better, cheaper and 3G, pepped up with HSDPA, as a viable alternative for fixed broadband access services like ADSL.

Cut the Cord, Talk to Me

The higher capacity and lower unit price of WCDMA provides a possibility for a fixed mobile substitution strategy. This is especially interesting for pure mobile players in markets with fixed telecom services at a relative high price or poor availability.

'The opportunity around fixed line substitution is happening now. The initiatives around bundled minutes – valuable plans with bundled minutes included – has started to change consumer's usage patterns and business users patterns. So I think that our market historically has seen high handset subsidisation, contracts and artificially high tariffs. As a result, relatively low usage, and over the last 12 months you have seen that turned on its head'.

The higher bandwidth and terminal capabilities of 3G create new mobile data service opportunities in the consumer and enterprise market. Operators with an existing GPRS service offering present 3G as an evolution; more speed, cheaper, better experience. For other operators it means the creation of a mobile data value chain and building a mobile data service portfolio. One executive about his 3G positioning:

'I'll say that 3G services have nothing to do with 3G technology. What our customers will see when we launch 3G technology, is that we will introduce some of the services that is possible on a high capacity network, primarily based on video streaming. We will also complete this portfolio with interactive location-based services. So, these will be new services where you need 3G technology. But we're not going to say, "Subscribe to 3G." We're going to say, "If you want video streaming, come to us." Customers don't care a hoot about technology!'

Here lies another controversy between operators; how to brand 3G services. The executive above clearly plays down 3G as a technology and does not use the term in his marketing.

Other operators put the term 3G and their new network at the core of their market communication. Although using technology as a marketing vehicle has been criticised in the past (remember WAP?), this strategy of branding the 3G technology network might actually make sense as a way to fight growing commoditisation of transport and distribution channels.

Cut the Cord, Mail to Me

An interesting opportunity for 3G is offered in countries like Philippines and Indonesia that have a poor fixed telecommunication infrastructure. One executive about his broadband ADSL connection:

'And if you look at DSL and cable, (the take-up) it is even lower. I pay US$30–40 a month to get broadband access, or so they tell me. I say otherwise, because I only get 100kbps here when I download, though I'm used to getting 512kbps or 1.5Mbps back home. It is a humbling experience.'

At the same time, it is typical for these countries that the number of e-mail addresses outnumbers PCs by a significant factor, indicating a pent up demand. 3G, in combination with HSDPA, provides a viable alternative and a business opportunity for mobile players. In the words of one executive in South East Asia: *'This leaves a service gap in the industry. Looking into the future, 3 to 5 years down, then the mobile data service becomes a viable alternative to the fixed data service. That is when you talk about 3G becoming a real alternative.'* But even in countries with a more developed fixed infrastructure there will be segments in the market that will be more than satisfied with the speed and coverage provided by 3.5G. Operators with established 3G networks will have relatively low incremental cost servicing this segment.

Approaches

NTT is currently developing:

(1) Optical ring system an optical ring system can be built simply by interconnecting nodes in a ring (Fig. 7.2). Therefore, a large-capacity optical network can be built economically. Since the remote control of nodes is easier in a ring network than in conventional networks, operations and maintenance are also easier. NTT is developing not only node systems for the ring network but also the key devices, such as switches, used in the node systems.

(2) Robust authentication platform and time stamping: An integrated authentication and time stamping mechanism achieves secure transactions by checking each access attempt against the caller ID, which is associated with the line that the user is using, and by attaching

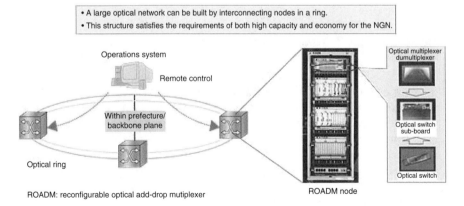

Figure 7.2 Optical ring system supporting the NGN

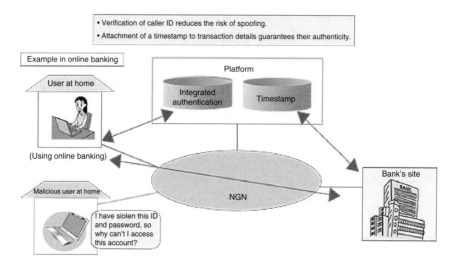

Figure 7.3 Safe and secure authentication system

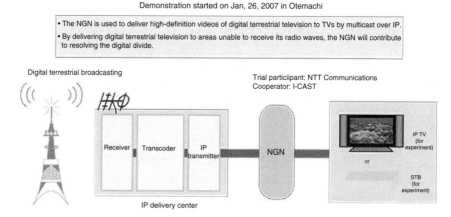

Figure 7.4 Retransmission of digital terrestrial television over IP (Source: see footnote 9)

timestamps to transaction details (Fig. 7.3). When the user at home attempts to access an online banking service, her caller ID is automatically verified so that she can conduct transactions, such as bank transfers, easily and securely. If someone tries to access the bank account from a place other than the authentic user's home, the caller ID does not pass the check and the network sends back a request for additional authentication information. This prevents fraudulent use of stolen passwords. Moreover, since bank transaction details have timestamps, any falsification of transactions can be identified, so only authentic transactions are processed and recorded.

Emergence of Global Network Service Providers[7]

The increasing range of services demanded by consumers has led to complex telecom solutions, with no one global network service provider owning every aspect of the solution. Ownership of network assets alone is not the sole criteria for the evaluation of Global Network Service Providers by enterprises needing their services. Instead, the capability to successfully integrated network elements from different sources and the ability to deliver good overall end to end services are the critical elements to be taken into account for evaluation. The need to factor these criteria would become increasingly important as the demand for global wireless services continues to grow. However, the supply to meet this Global Network Services is limited because the cellular market is just beginning to mature for them to take on multi-national ventures.

Global Network Service Providers market remains competitive, but with no real public price lists – which makes competitive bidding process as the best way to determine market pricing with benchmarking as the next best approach.

WHO ARE THE LEADING GLOBAL NETWORK SERVICE PROVIDERS?

According to Gartner they are the same as from five years ago, viz., AT &T, BT Global Services, Orange Business Services and Verizon Business. International operations of the Global Carriers alone would not be able to make the GNSP profitable, but should be coupled with strong domestic revenue generation. This has led some of the GNSPs, e.g. BT, Cable and Wireless, and Orange Business Services to move their operational support to lower cost locations and others to rely more on global service intermediaries for support.

Global Study – Findings, Need for Continuous Innovation

Over the next two years, two thirds of corporate CEOs who were interviewed indicated that they are going to need to make fundamental changes to their business. The reasons, they say, are many: intensified competition,

---

[7]Material for this section has been taken from the publication of Gartner Inc. Rickard, Neil, Paulak, Eric. 8 August 2007. Magic Quadrant for Global Network Service Providers. Gartner RAS Core Research Note G00150113. RA 1002152008. Link to web-site: http://www.corp.att.com/awards/docs/gartner_gns2007.pdf

escalating customer expectations, and unexpected market shifts. For many, you can add to that list workforce issues, technological advances, regulatory concerns, and globalisation. Yet fewer than half of CEOs think their organisations have handled such changes with much success in the past. These and other findings are in *Expanding the Innovation Horizon*, the 2006 IBM Global CEO Study reporting on the agenda of CEOs in the next few years. The results are based on interviews conducted recently by IBM and the *Economist Intelligence Unit* (EIU) with 765 chief executives from around the world. (It was the second global CEO study conducted by these partners; in 2004, the first Global CEO Study reported on interviews with 456 CEOs.)[8]

Among the study's findings were:

- *Business model innovation is becoming the new strategic differentiator.* 'The business model we choose will determine the success or failure of our strategy,' one study participant said. In contrast to the findings of the 2004 survey, innovation in the enterprise's business model garnered nearly as much attention as innovation in a company's core processes and functions.
- *Business model innovation can pay off.* In the financial analysis for the study, companies that have grown their operating margins faster than their competitors were putting twice as much emphasis on business model innovation as underperformers.
- *Innovation doesn't need a badge to get in.* CEOs said their company's employees were the most significant source for innovative ideas. But ranking close behind employees were business partners and customers – indicating that two out of the three top sources for the best ideas now lie outside the enterprise.
- *Collaboration can pay off.* The financial analysis explains why CEOs are more eager to partner and engage with other organisations than ever before. Companies with higher revenue growth reported using external sources significantly more than the slower growers. As one CEO said: *'If you think you have all the answers internally, you are wrong.'*

SEGMENTATION OF THE TELECOMMUNICATIONS INDUSTRY – THREE SEGMENTS

What is a mobile telecommunications operator? Today the most common answer is something like *'a company that owns and operates a mobile network and creates and provides services to customers over that network.'* However,

[8]Source: http://www.ibm.com/ibm/ideasfromibm/us/enterprise/mar27/ceo_study.html

there is a common view among many executives in the industry that this will change in the next years. Some executives interviewed even see dramatic changes coming to their telecom industry.

> *'Certainly, within the next 5 years, you will not see the industry as you have today. You will see a complete meltdown of the different players and the different sectors. You will not talk about wireless, wire line, cable or broadcasting operators.'*

The general picture is that three types of companies or 'blocks' will evolve with distinctly different business drivers and characteristics; the content and media block; the transport and distribution block and the service packaging and provisioning block.

Content and Media

The content and media companies own digitised content that they need to delivering over a multitude of carrier technologies. But in the end, they serve the same market. With mobile operators looking increasingly at the service provider role, there is less and less space for independent content aggregators. That is at least the opinion of some of the interviewees: *'Over a transition period, I believe that there is a role for content aggregators and brokers. But I think that is a very short period of time.'* Or another comment:

> *'In the service provider industry, the entrance barrier is much lower in the beginning. With competition getting fiercer and end users getting more critical, the operators' demand to service providers is getting tougher. Only the reputed service providers will be selected. I can forecast that there will be only several big national or international service providers to offer content and application for all operators.'*

Having said all that, it will take time for these changes to take place, and at least two interviewed CEOs had a quite different opinion. Legacy network and the ability to change will dictate the speed with which operators can adapt to their changing environment.

The transport and distribution block in the middle will be under increasing price pressure. This is for a number of reasons. First of all, there is the trend where access and transport networks are separated from the intelligent core and service networks. This trend is driven by standardisation of a layered network architecture, especially in IP Multimedia Subsystem (IMS) standards, and implemented in new hardware and software that becomes available to operators. A common way of describing this transition is to talk about moving from a 'sausage' model with vertically integrated, stand alone networks, to a 'hamburger' model with horizontal layers. This technology trend opens the way to services that are (access) network independent and that can be run over a choice of networks. And it is not just that these 'converged' services become available; customers expect them.

The transport and distribution function in the chain has to find ways to bring down cost by streamlining operations and creating economies of scale. Again operators have different solutions for the same issue. The most rigorous step is to outsource network related processes, sometimes to vendors like Ericsson. Outsourcing of network rollout has been common for many years but recently operators started contemplating outsourcing of more complex functions such as network planning and optimisation, maintenance and operations. One of the interviewed operators had actually taken that step. A less far-reaching step is to share cost with competitors in non-differentiating areas such as base station sites. As one CEO explained:

> *'The sharing of costs and reduction of costs that I talked about in terms of a maturing marketplace will still be very relevant going forward, so the infrastructure sharing agreements you have seen in this marketplace are just a huge flag waving about the changing of behavior between competitors. There is more of a realisation that no-one is going away, everyone's got to make money, so let's work with each other in areas that are non-differentiating to try and reduce cost.'*

Service Packaging and Provisioning

The service packaging and provisioning block has the interface with the customer. Their challenge is to identify customer needs and translate this into a constantly changing set of services or *'customer propositions'*, as one CEO put it, that customers are willing to pay for.

The consequence of all this is that it will be increasingly difficult for operators to differentiate themselves with their (access) networks. The content and media companies as well as the service providers will have a choice of channels to bring their services to the market and customers will see increasingly less value in the transport and distribution function. The 'pipe' will become a commodity with declining revenues as a consequence. This leaves current mobile operators with the question what type of company they want to be in the future. One CEO gives the following hint:

> *'You might see an operator in the future, like myself, where I put my focus on the market. I may own some part of the delivery technologies because I have the capability of running a mobile network – or perhaps not, as I may not be limiting my services to the mobile network. So, I may use transport technologies from other network providers to deliver a complete service package.'*

So, given that the mobile marketplace will change significantly over the next years (all but two interviewed executives made some sort of statements on rearranging of the mobile value chain), how do mobile operators prepare themselves? When it comes to where the future business value is, most are in agreement:

> 'The key is to come to the market. The guy sitting here in between is becoming more and more squeezed. This space is becoming more and more uninteresting, because there is so little you can do when you're developing the business.'

Furthermore, access to content will not be a differentiator:

> 'The model that goes for content as a differentiator is a fundamentally flawed model as everyone is trying to open up this marketplace to get more people using more content. To have narrow exclusivity on content is more expensive, and by definition, starts to reduce the number of people using it. It is going to slow the marketplace down. I personally believe the way you will see content as a differentiator is in terms of how it is presented, not in terms of whether you can access this piece of content or not.'

Current operators are in fact well positioned to take this role of service provider cum marketing force for several reasons. First of all they have an established (billing) relation with their customers, some of them have worked hard on their brand recognition and they are sitting on a gold treasure of information about their customer. As one confident CEO puts it:

> 'And I think this industry have to capture that area. We can do much better marketing, much more direct, much more relevant because we have better CRM. Our cost structure should be lower, provide our vendor doesn't overcharge us. We could do it!'

On the question how to get there, different answers were given during the interviews.

> 'There is a long line of evolution that we might see. In that context, when that day comes, it might be that this legacy network that we have with us is not our focus anymore, as we move into a service package. We're not there yet. It may be a possibility in 5 to 10 years from now.'

Where executives disagreed with the vision of a market aligning around three blocks the main argument was that, in order to differentiate, the operator has to have end-to-end control of the service and customer experience. Leaving parts to others will create an uncertainty in the supply chain that is not acceptable. In the end all links in the chain will be needed to provide services that customers are willing to pay for, and a successful transport and distribution company might be more profitable than a not so successful service provider. As much as choosing the right strategy, it will be equally important to choose a clear strategy and execute on it.

CREATION OF SHARED GLOBAL KNOWLEDGE NETWORKS

Wikipedia is a free encyclopedia that allows anyone, not just selected experts, to contribute to its content over the Internet (Fig. 7.5). This project was started in 2001 in the USA. The Japanese language version was also started the same year and today has more than 320 000 entries, which means that it is more comprehensive than any ordinary paper-based encyclopedia. We expect that the Internet encyclopedia will continue to improve and grow through the sharing of knowledge by many people.

World Bank – Transforming Development Through Information and Communication Revolution

Today we have a unique tool at our disposal to enable involvement of all, on a scale undreamed of even several years ago. The information and communications revolution will transform development, as we know it. This revolution promises a historic opportunity to redraw the global economy through broad and equal access to knowledge and information;

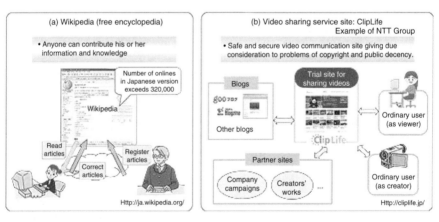

Figure 7.5 Examples of overcoming the unequal distribution of knowledge[9]

[9]Norio Wada. The Transformation and Future Direction of Communications. President and CEO. https://www.ntt-review.jp/archive/ntttechnical.php?contents=ntr200704sf1.html Last accessed on 12 February 2008.

through enhanced empowerment and inclusion of local communities; and through economic growth, jobs, and improved access to basic services. And so, over the last five years, we have been focusing on how we can harness the power of information and communications technology and of knowledge to accelerate development.

We are working with governments to foster policy, regulatory, and network readiness, through our analytical and advisory work and through our grant facility, *info* Dev.

We are linking development leaders globally through our Global Development Learning Network, which provides training and creates broad learning communities.

We are connecting students and teachers in secondary schools in developing countries with their counterparts in industrialised countries through our World Links for Development program.

We are using information and communications technology to create a 'university without walls' and to link sub-Saharan African countries directly with global academic faculty and learning resources through the African Virtual University.

Through the Global Development Gateway and the Global Development Network, we are promoting the generation and sharing of knowledge. We are supporting knowledge networking, global research, and communities of practice from the grass roots up.

And finally, we are developing many practical applications being used by poor communities worldwide to bring them knowledge in local languages, to build communities, to create businesses, to assist in medical treatment and to link them with each other and with the world.

The information and communications revolution offers us an unprecedented opportunity to make empowerment and participation a reality. And poor people across the world are demanding action. In response to our *Voices of the Poor* study, many groups have asked us, as a key priority, for increased access to information and communications technology.

We must work toward the day when, through the Internet, through distance learning, through cellular phones and wind-up radios, the village elder or the aspiring student will have access to the same information as the finance minister.

Communications technology gives us the tool for true participation. This is leveling the playing field. This is real equity.

We Have a Historic Opportunity

This new world, our greater understanding, a wiser development community and a changing international institutional environment mean that through working together, doing development differently, and giving voice to the voiceless, we now have a chance to make the next decade one of real delivery in the fight against poverty. The opportunities and promise of a global economy, the information age, and life-saving and productivity-enhancing technologies are all ours to seize. We must work together to harness the benefits of globalisation to deliver prosperity to the many, not just the few (Wolfensohn, 2000).[10]

TOP OR FLOP? THE SPONTANEOUS REACTION OF MOBILE EXECUTIVES TO 'HOT' TECHNOLOGY

We interviewed number of c-suite executives of leading mobile operators in the Asia-Pacific region on their take of the market in the next 3–5 years, focusing on how this will impact their company and what strategies they have. At the end of each interview, we showed the executive a list of technologies that we picked from the Cannes 2005 3GSM conference press releases. Our question was short and to the point:

> *'Is this going to be a success in the market or not? Top or Flop.'*

The spontaneous answers paint a picture of three groups; Tops, Flops and Undecided.

Tops:

Mobile Marketing

One definite Top was mobile marketing; enterprises using the mobile channel to reach (potential) customers, either for promotional or relationship management purposes. In the words of one enthusiastic executive:

> *'Especially with 3G coming on your little screen. There are lots of content that can be delivered to your mobile devices. That gives opportunity to deliver marketing. It's like cable TV, putting advertising in between programmes.'*

However, concerns about personal privacy and the regulatory backslash that might follow were voiced as well:

[10] Excerpt from speech: 'Building an Equitable World' by James D. Wolfensohn, Former President of the World Bank to the Board of Governors, Annual Meeting, Prague, Czech Republic, September 2000. Link: http://web.worldbank.org/ WBSITE/EXTERNAL/COUNTRIES/ECAEXT/EUEINPEXTN/0,contentMDK: 20025477~menuPK: 590789~pagePK: 2865066~piPK: 2865079~theSitePK: 590766,00.html

'The various government authorities here are concerned here with the proliferation of these types of services affecting the privacy of users. So that will be a market-by-market phenomenon.'

Mobile Wallet

Another clear Top was the mobile wallet, the concept of using the mobile device to make financial transactions, either via a pre-paid debit card set-up or via the regular billing process of the operator. In the eyes of some executives there is a strong link with the previous topic, mobile marketing:

'It goes in line with our marketing approach, if you want to be a marketing platform for other industries, a way of effective payment is important. You can be a facilitator for credit card verification or that sort of thing.'

It is also clear that a mobile wallet concept is a particularly difficult one to implement:

'I do believe in mobile wallets, but there is a gap between the concept and the execution in your market and get the other industry players to come and play. If you don't do that you don't get anywhere.'

In particular, the banks are sometimes difficult to get on board:

'One of the biggest impediments has been the participation of the banking industry, because they have been reluctant for telcos to be involved in what they call financial switching. What is happening now is that the mobile carriers are co-operating to persuade the banks that this has to happen.'

Flops:

Location Based Services

A service that scored low on the executive hit list was Location Based Services. It is seen as a niche application for certain regions and users: *'Maybe to locate children or pets'*. With five Flop votes, one undecided and two Tops, location based services do have some supporters. One supportive executive remarked:

'So I think location services are more immediate, within the next twelve months. It has a good future, it's effectively a complementary service to enable customers to get in contact and actually perform other services and we are hoping that it is a complementary service that generates more calling revenue.'

Push to Talk

Another service that received a Flop judgment from our panel was Push to Talk. Successful as it might be in the States, the Asia Pacific sector does not warm up to the proposition of walkie-talkie on the mobile. As one skeptical executive says:

> 'I think the success in the US for Nextel is peculiar to their environment – for that market, at that time. Even others tried to do it in the States and it didn't work. Technically it can be done. But when we look at customer behaviour and characteristics, I ask, why?'

The topic of customer behavior refers to talking loud in public places (with everybody listening in on the reply) and the practicality of making this type of calls in a metropolitan noisy environment. In countries with low regular voice tariffs the Push to Talk proposition was not seen as attractive enough to steal away regular voice traffic.

WiMax Technology

The last clear Flop in the list was the much-discussed WiMax technology. Although WiMax suitable spectrum was auctioned in Singapore in May to six different buyers for a total of around US$5.8m, none of the executives we interviewed actually believed in the technology as bringing a superior proposition to the market. As one executive explained:

> 'There is a lot of information on how cheap these technologies are. We're looking at WiMax and Wireless IP. We are looking at applications that we can use and what if we don't use them. They all try to offer that holy grail of full mobility. Technologically, it will take sometime. There will be big battle on standardisation and following that you need the devices to support it. What is the architecture? And whether they can offer full mobility at the same level of cost structure.'

HSDPA received a better rating, probably because it is a more sensible step for operators with a 3G network in place to upgrade with this technology.

Undecided:

Mobile TV

The applications that received an undecided vote from our executives were video telephony and mobile TV. Arguments against video telephony, a standard feature and differentiator for most 3G business models, were that it is a niche application, that it needs a certain phone penetration to take off (no use trying it if the other party does not have a phone with video call capabilities) and that people might not like to be seen at any time and any place. On the other hand it is an appealing service with great marketing potential. The mixed feelings were well described by one CEO:

'It won't be strong, it won't be top but I can't see it being a flop. I don't believe it will be a mass-market application in the next 3 years. It's niche but it's gold as a niche application. If you are video calling overseas or interstate, its fantastic but you'll use it a small percentage of the time. It doesn't fall into either (top or flop) but I know we sell product because of it. It positions 3G in a superior light to GSM so I would not want our product to not have it, but I don't believe it drives dramatic incremental revenue going forward.'

The proposition for mobile TV is a tricky one that requires careful planning. Do you imagine people watching one hour of soap on their phone? Probably not! Can you imagine the value of streaming live, breaking news? Yes. It seems that mobile TV is more for the 'info-snacking' user. Key success factors mentioned by our interviewees were: interactivity, creativity to take it beyond today's TV format, short clips, life feeds. One executive summed it up for us:

'It may be the next service that captures the imagination of the user. The question is what is the revenue model? Not sure. You have to define what that means. The idea that you stream 90 minutes of AC Milan versus Inter Milan on your mobile – I just don't buy it. I think there is unquestionably scope for some level of interaction with TV shows but I'm not sure what that means. If you were out and about when 9/11 broke or when the US invaded Iraq and you got alerted and could be streamed in, that's good stuff.'

CONCLUSION

As the telecom industry is famous for both hyping and surprising it will be interesting to see how our hit list and business models develops over time. We like to come back to you with an update in a year's time, to talk again about WiMax, HSDPA, Mobile TV and Wallets with some more hindsight and see how wrong we were. As the weatherman says, forecasting is darn difficult, especially if it is the future.

8

Internet in Rural Areas – Emerging Business Models and Opportunities in Developing Countries*

OVERVIEW

The provision of information, knowledge and Internet based services are increasingly being recognized as essential elements to promote economic growth and reduce poverty in rural areas. Recent advances in information and communication technologies (ICTs), especially computers, electronic networks and Internet based applications along with traditional forms of communication media (such as radio, telephone, television, etc.) have the potential to be used as a means to empower poor people, enhance skills, increase productivity, share market information on prices and opportunities, improve participatory decision-making processes, governance at all levels and to provide effective and efficient delivery of government services. Internet based services provide both immense opportunities and challenges in meeting these increasing demands and in improving the lives of the rural population. This chapter focuses on the role of the Internet in rural areas and presents a few emerging business models and opportunities in developing countries.

*The views and opinions expressed in this chapter are the individual author's own and should not be attributed to the World Bank, its management, its Board of Directors or the countries they represent.

Business Models for Sustainable Telecoms Growth in Developing Economies
S. Kaul, F. Ali, S. Janakiram and B. Wattenström
© 2008 S. Kaul, F. Ali, S. Janakiram and B. Wattenström

CURRENT STATUS OF INTERNET IN RURAL AREAS

Rural areas are confronted with low levels of income, poor technical skills, difficult geographical terrain, deep-rooted cultural practices and low levels of affordability. Despite significant advances in technology, more than two thirds of the world's population remains without access to communication facilities. In Africa, only 2.6% of the population has Internet access and 3.7% have access to either a fixed line or a mobile telephone.[1] This situation could change if an enabling business, institutional and policy environment is created for the introduction of emerging telecommunication technologies which offer cost effective means for Internet access for the rural population. For example, wireless technologies offer opportunities to bring low-cost, accessible information and communication technologies to those who have been left behind. These include WiFi (Wireless Fidelity), mesh networks, WiMAX (Worldwide Interoperability for Microwave Access), Orthogonal Frequency Division Multiplexing (OFDM), and 3G mobile services.

Internet Access

Internet access in rural areas in both the developed and developing world is unequally distributed in relation to those in urban and peri-urban areas. This Internet divide is being reduced at a much slower pace in developing countries than in the developed countries. Figure 8.1 shows the costs of broadband for different regions of the world.

Prices paid by users for broadband services continue to be highest in the Africa region by about four to eight times as compared to users in Oceania or in the Americas. This differential could largely be attributed to lack of competition in the provision of broadband services in many parts of the world (see figure 8.1). Cost of access is often the main reason why households, especially in rural areas, constrain Internet access. One study indicated that at an income level of US$44 000, cost of Internet access ceased to be a factor and were overridden by whether the household had any interest. Other reasons include inadequate computer capacity and inability to access the Internet outside the home.

A number of international studies indicate that a range of social factors influence the existence of a digital divides. For example, research in the

---

[1]Comminit. 2007. http://www.comminit.com/trends/ctrends2006/trends-296.html last accessed on 11 August 2007.

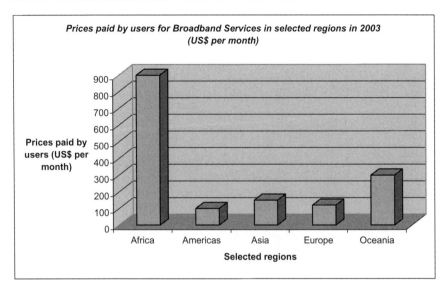

Figure 8.1 The main problem – Cost of access[2]

United States suggests that sharp inequalities exist in Internet access according to income, education, age, race and ethnicity. The poor, minorities (particularly the black population), poorly qualified and single mother households are much less likely to be online than their younger, more educated and upper income professional counterpart Americans.[2] The Organisation for Economic Cooperation and Development (OECD) has documented similar patterns of stratification among the Internet population across other developed countries. For example, for every US$10 000 increase in income, the likelihood of a household owning a computer increases by 7%. With respect to Internet access, households with higher incomes are far more likely to be connected, compared to those with lower incomes. Cross-national statistical evidence also suggests that ethnicity and race are determinants of participation online, even when corrected for income.

[2] World Bank. 2007. Connecting Sub-Sahara Africa: A World Bank Strategy for ICT Sector Development. Global Information and Communication Technology Department (GICT). Working Paper, No. 51. World Bank, Washington DC.

Table 8.1 Internet access, quality, affordability and applications, 2005

1. Region	Quality			Affordability			Applications		
	2. Internet access (users per 1000 persons)	3. Broadband subscribers (per 1000 persons)	4. Int'l Internet bandwidth (bits per person)	5. Price basket of Residential fixed line ($ per month)	6. Price basket of Mobile telephone service ($ per month)	7. Price basket of Internet service ($ per month)	8. Sector expenditure (% of GDP)	9. E-Govt Index	10. Secure internet servers (per million persons)
1. East Asia and Pacific	89	25.9	97	5.9	5.0	10.7	5.3	0.38	1.0
2. Europe and Central Asia	190	20.9	211	9.5	11.8	12.2	5.1	0.51	13.1
3. Latin America and Caribbean	156	16.5	161	10.0	9.4	25.8	5.9	0.48	12.0
4. Middle East & North Africa	89	0.5	9	7.3	6.3	11.8	3.1	0.33	0.7
5. South Asia	79	1.0	18	5.1	2.4	8.1	5.7	0.29	0.6
6. Sub-Saharan Africa	29	0.0	2	14.0	12.3	45.3	–	0.25	2.0
7. Middle income countries	115	22.6	92	9.7	10.1	17.0	5.4	0.46	5.2
8. Europe (EMU)	439	134.7	5784	29.0	20.9	20.4	5.4	0.71	185.0
9. High Income	579	163.2	4537	27.6	17.8	19.9	7.2	0.77	444.4

Notes:
1. The regions presented in table 8.1 correspond to groupings of countries used by the World Bank and may differ from common geographical usage.
East Asia and Pacific consists of the following countries: American Samao, Cambodia, China, Fiji, Indonesia, Kiribati, Korea, Democratic Republic, Lao PDR, Malaysia, Marshall Islands, Micronesia, Fed. Sts, Mongolia, Mynamar, Northern Mariana Islands, Palau, Papua New Guinea, Philippines, Samoa, Solomon Islands, Thailand, Timor-Leste, Tonga, Vanatu and Vietnam.

Europe and Central Asia consists of the following countries: Albania, Armenia, Azerbaijan, Belarus, Bosnia and Herzogovina, Bulgaria, Croatia, Czech Republic, Estonia, Georgia, Hungary, Kazakhstan, Kyrgyz Republic, Latvia, Lithunia, Macedonia, FYR, Moldova, Poland, Romania, Russian Federation, Serbia and Montenegro, Slovak Republic, Tajikistan, Turkey, Turkmenistan, Ukraine and Uzbekistan.

Latin America and the Carribbean consists of the following countries: Argentina, Barbados, Belize, Bolivia, Brazil, Chile, Colombia, Costa Rica, Cuba, Dominica, Dominican Republic, Ecuador, El Salvador, Grenada, Guatemala, Guyana, Haiti, Honduras, Jamaica, Mexico, Nicaragua, Panama, Paraguay, Peru, St.Kitts and Nevis, St. Lucia, St. Vincent and the Grenadines, Suriname, Trinidad and Tobago, Uruguay, Venezuela, RB.

Middle East and North Africa consists of the following countries: Algeria, Djbouti, Egypt, Arab Republic., Iran, Islamic Republic, Iraq, Jordan, Lebanon, Libya, Morocco, Oman, Syrian Arab Republic, Tunisia, West Bank and Gaza, Yemen Republic.

South Asia consists of the following countries: Afghanistan, Bangladesh, Bhutan, India, Maldives, Nepal, Pakistan and Sri Lanka.

Sub-Saharan Africa consists of the following countries: Angola, Benin, Botswana, Burkina Faso, Burundi, Cameroon, Cape Verde, Central African Republic, Chad, Comoros, Congo, Democratic Republic, Congo, Rep, Cote d'Ivoire, Equatorial Guinea, Eritrea, Ethiopia, Gabon, The Gambia, Ghana, Guinea, Guinea-Bissau, Kenya, Lesotho, Liberia, Madagascar, Malawi, Mali, Mauritania, Mauritius, Mayotte, Mozambique, Namibia, Niger, Nigeria, Rwanda, Sao Tome and Principe, Senegal, Seychelles, Sierra Leone, Somalia, South Africa, Sudan, Swaziland, Tanzania, Togo, Uganda, Zambia and Zimbabwe.

Middle Income Countries are those with a Gross National Income (GNI) per capita of more than US$875 but less than US$10726. Europe includes the member states of the Economic and Monetary Union (EMU) of the European Union which have adopted the Euro as their currency, viz., Austria, Belgium, Finland, France, Germany, Greece, Ireland, Italy, Luxemburg, Netherlands, Portugal, Slovenia and Spain.

High Income countries are those which have a Gross National Income (GNI) per capita of US$10726 or more.

2. *Internet users* are people with access to the worldwide network. (International Telecommunication Union).

3. *Broadband subscribers* are persons with a digital subscriber line, cable modem, or other high-speed technology connection to the Internet. Reporting countries may have different definitions of broadband, so data may not be strictly comparable across the different countries.[3]

4. *International Internet bandwidth* is the connected capacity of international connections between countries for transmitting Internet traffic. (International Telecommunication Union).

5. *Price basket for residential fixed line* is calculated as one-fifth of the installation charge, the monthly subscription charge, and the cost of local calls (15 peak and 15 off peak calls of three minutes each. (International Telecommunication Union and the World Bank).

6. *Price basket for mobile telephone service* is calculated based on the prepaid price of 25 calls per month spread over the same mobile network, other mobile networks, and mobile to fixed calls and during the peak, off peak and weekend times. The basket also includes 30 text messages per month. (International Telecommunication Union).

7. *Price basket for Internet service* is calculated based on the cheapest available tariff for accessing the Internet 20 hours a month (10 hours peak and 10 hours off peak). The basket does not include the telephone line rental but does include telephone usage charges, if applicable. Data are compiled in the national currency and converted to US dollars using the annual average exchange rate. (International Telecommunication Union).

8. *Sector expenditure* includes computer hardware (computers, storage devices, printers, and other peripherals); computer software (operating systems, programming tools, utilities, applications, and internal software development); computer services (information technology consulting, computer and network systems integration, web hosting, data processing services, and other services); and communication services (voice and data communication services) and wired and wireless communications equipment. (World Information Technology and Services Alliance).

9. *E-government readiness index* is based on a five-stage ascending model that builds on a government's previous level of sophistication of online presence. The stages are emerging, enhanced, interactive, transactional, and networked. Countries are scored on a scale of 0 to 1, with 1 indicating most ready according to the specific products and services they provide. (United Nations Department of Economic and Social Affairs, United Nations Online Network in Public Administration and Finance).

10. *Secure internet servers* are the number of servers using encryption technology for internet transactions. (Netcraft).

[3] International Telecommunication Union (ITU), 2007.

EXAMPLES OF INTERNET USE FOR RURAL DEVELOPMENT

Brazil: Free Internet Access to Empower Indians[4] to Protect Rain Forest

The Brazilian government has launched a program for providing Internet connectivity in rural telecentres for the use of the Indian tribes who reside in remote areas of the Amazon rain forests. The Forest People's Network was established to strengthen monitoring, protection and education on environmental issues. The expected outcome was an improvement in the enforcement mechanisms of the government against illegal mining and logging through empowerment of the Indian tribes. The Government provided free access to Internet services via satellite to about 150 communities, some of whom could be reached only by boats. Francisco Costa of the Environment Ministry of Brazil stated that *'The goal was to encourage those peoples to join the public powers in the environmental management of the country.'* As a prerequisite, in order to obtain free Internet access, the city and state governments were required first to install telecentres with computers in selected areas, including indigenous lands, following which the government would provide the necessary satellite connections.The network helped to increase the collaborative effort between the government and the Indians.

Egypt

The agricultural sector in Egypt is increasingly becoming a fully private sector operating in a market and export oriented economy. This has been achieved through major agricultural reforms initiated in the 1980s and continued through the 1990s, such as price and trade liberalisation in input and output markets and the elimination of land use controls for most crops. Despite these reforms, farmers continue to face the following constraints: (i) the system of market integration between the farm and retail markets has changed little since the beginning of the reform programme; (ii) linkages between agricultural research, extension, education and farmers need further strengthening; (iii) inadequate information, knowledge and support for improving exports. These constraints prevent the farmers from reaching their full potential, especially in cotton and horticultural crops, in terms of increased agricultural productivity, obtaining their fair share of profits and taking advantage of export opportunities in the regional and western markets. Addressing these constraints requires

[4] Rights and resources Blog. 2007. http://rightsandresources.org/blog/labels/Fo rest%20People's%20Network.html last accessed on 14 September 2007.

the provision of reliable, timely and relevant information and knowledge and business services aimed at meeting the changing needs of the farmers and markets. Recent advances in information and communication technologies have made it possible to provide these services in a cost effective and sustainable manner. Over the last few years, the government of Egypt has made significant progress in the use of ICT and introducing internet based applications to improve rural livelihoods and promote economic growth. Following are some examples being carried out.

Virtual Extension and Research Communication Network (VERCON)

An Internet based Virtual Extension and Research Communication Network (VERCON)[5] has been established by the Ministry of Agriculture, in cooperation with FAO. VERCON supports research/extension interaction and flow of information. It is a cooperative effort between the Central Lab of Agricultural Expert System (CLAES), which is the hosting institution and the Agricultural Extension and Rural Development Research Institute (AERDRI), which is responsible for the extension aspects. VERCON is being used by farmers to access information and knowledge from a variety of sources and to seek advice from agricultural specialists on agricultural practices, pest management, crop diversification, artificial insemination, etc. The VERCON pilot has successfully demonstrated that the system and its users can improve linkages between research and extension in eight pilot nodes.[6] It has the potential to form the basis for the creation of a national electronic agricultural knowledge and information network. Such a network could improve and further strengthen the generation, flow, sharing and collaborative use of agricultural knowledge and information.

Rural and Agricultural Development Communication Network in Egypt (RADCON)

The main objective of RADCON is to improve the communication between extension, research, private and public sectors and institutions involved

---

[5] The VERCON pilot initiative is funded by the Ministry of Agriculture and Land Reclamation (Egypt) in partnership with the Food and Agriculture Organization of the United Nations (FAO) through a Technical Cooperation Project agreement, and supported by FAO's Research, Extension and Training Division together with FAO's World Agricultural Information Centre (WAICENT). It is available in Arabic at www.vercon.sci.eg

[6] Sallam, S.M. and Kassem, M.H. 2002. Virtual Extension and Research Communication Network (VERCON) in Egypt – Linking Extension and Research – Agricultural Extension Aspects.

in rural and agricultural development for the benefit of farmers and agrarian businesses at rural and village levels. It would provide a sustainable operational dynamic information and communication system that responds to the stakeholders requirements including resource poor communities. Fifty resource poor communities are being identified and enabled to participate in the RADCON activities and benefit of its knowledge and information resources. New modules are being developed, consisting of marketing, community services, women's corner, environment and youth job opportunities modules. The server in CLAES uses a Windows 2000 Platform, MS SQL database engine and bandwidth connectivity to the Internet of 2 Mbps. All sites are connected to through the national backbone. A team of CLAES software engineers (20 to 25 persons) are currently developing all the applications of the RADCON system. RADCON is currently under implementation.

Ministry of Communication and Information Technology Initiatives

Since its establishment in 1999, the Ministry of Communications and Information Technology (MCIT) has been aggressively promoting the development of a networked economy and initiating reforms to foster competition in the communication sector in Egypt. For example, the Broadband initiative is being restructured in order to promote connectivity, especially to rural areas. The current charges are: (i) US$1/month for local loop unbundling, (ii) US$15/month for 256 kbps and US$26/month for 1 Mbp. Reforms in the telecommunication sector have resulted in the establishment of a state-of-the-art communication system, including three mobile phone companies servicing a subscriber base that has seen exponential growth over the last few years. Among the projects being implemented by the MCIT are SMART Village, ICT for Illiteracy Eradication Project, GIS system for land records and creation of a one-stop shop, etc.

Ghana: Tradenet – Web-Based Application for Market Price Information

Tradenet[7] was launched to address the problem of relevant and timely provision of market information for the African farmers, governments and service providers in the agricultural sector. Tradenet developed and

[7] This write-up is based on information provided in Tradenet's web-page: http://www.tradenet.biz last accessed on 15 September 2007.

field tested three different types of market information services over a period of five years, which were aimed at meeting the needs of farmers, traders, agro-processors and small scale farmers. They were:

- *Local market information*: which meets the specific information needs of small scale farmers and traders at the district level.
- *National market information service*: which provides regular updates of countrywide market status and is aimed at meeting the information needs of Government, national traders and food security agencies.
- *Regional market information service:* which provides information for the formal and informal traders who are involved with cross border trade of high volume staple commodities aimed at the needy. It consists of a set of software tools, databases and services for the use of those involved in the agricultural sector.

Over the last two years, Tradenet has established linkages with Foodnet, Techno serve, FAO, CGIAR and IFDC – institutions involved in the development of the agricultural sector. Tradenet is involved in over 300 markets collecting and disseminating price information. It allows anyone with a mobile phone to submit an offer, receive price quotes or to buy goods through Tradenet. Since it has only recently been launched, it is too early to assess the impact of Tradenet. Different institutions operate Tradenet in different countries. For example, USAID's Mistowa program operates Tradenet in 12 countries. CIAT operates Tradenet in Uganda.

The underlying business model is a licensing model. The user pays an annual fee which is used by Tradenet to provide the initial start-up, technical support, such as content uploads, replications and/or SMS integration or resolve access issues and host the application. Table 8.2 provides the basic costs associated with Tradenet:

India: Rural Kiosks Provide Information, Knowledge and Business Services in Andhra Pradesh

Rural poverty in Andhra Pradesh is multidimensional and complex. The government has not met the basic needs of the poor, resulting in poor

Table 8.2 Costs of services in Tradnet

Service/Fee	Cost (in US$)
1. Set up	2000
2. Annual fee	5000
3. SMS set up (optional)	500
4. Staging server (optional)	1000
5. On-site training (3 day)	1000

access to water and sanitation, poor health and malnutrition, continuing illiteracy, a lack of marketable skills, violence and crime. The lack of political freedom and participation of the poor has further alienated them. The situation has been exacerbated by slow growth in agriculture and limited diversification of the rural economy. In response, the Andhra Pradesh government developed its Vision 2020 for comprehensive human and economic development which mainly seeks to improve rural livelihoods for marginal and small-scale farmers, widen access to non-farm employment, improve community access to financial resources and to use recent advances in information and communications technology to reduce rural poverty.

Project Objectives and Description

The Andhra Pradesh Rural Poverty Reduction Project aimed to enable the rural poor, particularly the poorest of the poor, to improve their livelihoods and quality of life. Through the establishment of rural kiosks in selected project areas, information and communications technology were used to empower rural women, enhance skills, increase productivity, improve participatory decision-making, provide timely delivery of government services and build new links between segments of the rural population.

Rural kiosks – which have been in operation only since 2005 – nevertheless show good potential to help improve rural livelihoods. They are managed by Mandal Samakhya, a federation of self-help groups. Each kiosk caters to 1000–3000 households. The rural kiosk can be considered as an essential first step towards achieving a statewide networked economy, in which (1) the rural population has access to information for empowerment and development; (2) marginalised farmers can obtain information to move up the value chain, such as information on markets, tools to improve productivity, and best practices; (3) localised content and interfaces are developed to meet the needs of various types of households with varying degrees of literacy; (4) domestic and global markets are brought closer to those making products and artisans making handicrafts; (5) information on employment opportunities is available; (6) the opportunity exists to develop a fully e-literate state, in which at least one member of each family is computer proficient.

The criteria for setting up a rural kiosk include that they must be owned and managed by Mandal Samakhya and operated by self-help groups; they must have uninterrupted power supply and reliable telephone connectivity, easy access by rural households and close proximity to a bank (preferably walking distance). Based on these criteria, several steps were taken to set up rural kiosks. Two to four women from the self-help group were identified as having the aptitude and analytical abilities to work with local people to operate and manage the kiosks.

These women received hands-on training in computer applications and software for collecting utility bills. A suitable location in a busy area was selected – usually one or two rooms that were rented, donated or owned by the community – and rehabilitated with community support to ensure safety, power supply and customer services. After the installation of the computer and related accessories, the community was informed about the services being offered by rural kiosk. Support and involvement was solicited from educational foundations, establishments, agriculture departments, farmer organisations and others to provide content and learning materials. The kiosk was open for at least eight to ten hours each day to provide the advertised services and feedback was obtained from customers about ways to improve and increase the scope of services. The self-help group members who operated the kiosk were provided with incentives and encouragement to ensure that they would provide services demanded by customers.

Benefits and Impacts

The kiosks have benefited rural households in the following ways:

- Ease of paying electricity, telephone, and water bills – which is also a source of revenue for kiosk operators.
- Availability of market prices for agricultural products and commodities, which enables farmers to obtain better prices for their crops, avoid exploitation from intermediaries, and increase their incomes.
- Issuance of caste, income, and residence certificates which eliminates the need to travel long distances and reduces occasions for rent seeking by government officials.
- Provision of digital photography and lamination services.
- Provision of computer classes to increase marketable skills among the unemployed, especially the young people, who have been able to find jobs as a result.

The income generating services have helped make the kiosks financially viable. An investment cost ranging from about Rs175 000 (US$4000) to Rs247 000 (US$5600) per kiosk yields a net monthly profit ranging from Rs7000 (US$160) to Rs18,000 (US$400), depending on the number of households served and the range of fee- and demand-based services offered.

Figures 8.2 and 8.3 show notice boards of services offered at a rural Internet centre (in English and the local language Telugu) in Andhra Pradesh, India.

Figure 8.2

Figure 8.3

Lessons Learned and Issues for Wider Applicability

A number of important lessons have been learned, but the main lessons learned include:

- *Ownership and management* by motivated and enthusiastic women's self-help groups, eager to learn and provide client-oriented services, are essential for the sustainability of rural kiosks.
- *Continuous training* for the kiosk operators in applications of information and communications technology is important.
- *The selection of an appropriate location* is critical to increasing the client base. The location should offer potential to cater to the needs of local businesses, enterprises, temples and other potential customers
- *Strong marketing links between rural kiosks and procurement centers* are important. Community-owned procurement centres have been established by the project to serve the poor producers and consumers who are at the bottom of the value chain, lack marketing skills and are subject to exploitation by intermediaries. Linking the rural kiosk with the procurement centre are essential for ensuring that information and knowledge services provided by the kiosks are used to obtain the most competitive prices for primary producers and avoid exploitation by intermediaries.
- Over the last decade, the creation of a large statewide network of self-help groups and village organisations of the poor with links to commercial banks has laid the institutional foundations for motivating these groups to adopt innovative programmes, explore new frontiers and take the associated risks.
- *Strong leadership at the Andhra Pradesh State Government level*, with a vision to bring government services to citizens at minimal cost and a commitment to increase transparency, has been vital.

The success of the rural kiosks has provided the basis for establishing about 8600 rural kiosks in Andhra Pradesh through two large private operators. With the gradual addition of value-added services, rural kiosks could evolve into sustainable rural information, knowledge and services centres which would provide a combination of fee-based and free services to the rural community to improve livelihoods. This vision is consistent with the evolving national plan to create a network of village knowledge centres in India, in which the Internet would play a significant role.

N-Logue's Rural Connectivity Model

n-Logue Communications Ltd. has developed a for-profit business model designed to affordably meet the latent demand for rural connectivity. The company was incubated by the Telecommunications and Computer Networks (TeNeT) Group at the Indian Institute of Technology in Madras, as part of the group's mission to create appropriate and cost effective technology solutions designed for developing countries. The project was aimed at 'significantly enhancing the quality of life of every rural Indian' by setting up a profitable network of wirelessly-connected Internet kiosks in villages throughout India.

Business Model

To enable its rapid expansion, n-Logue employs a three-tier franchise business model that brings the delivery and management of Internet services closer to the end user. Each tier consists of independent, financially self-sustaining entrepreneurs operating interdependently with one another. At the top level is n-Logue, responsible for overall management of the network. The company facilitates relationships between its upstream partners – banks, governments, hardware and solution providers – and its business franchisees.

On the second tier are the Local Service Providers (LSPs), responsible for managing the project at the local level. In coordination with n-Logue, the LSP invests in and sets up an Access Center that provides last-mile access to subscribers in the project area. On the bottom tier of n-Logue's business model are the local entrepreneurs that are recruited by the LSP to invest in and set up Internet kiosks in their villages. The kiosk owners purchase the computer equipment through n-Logue, who also provide training, support and technical assistance. These locally-owned franchises offer a variety of Internet and computer-based services aimed at the rural market.

GRASSO: Self-Employment Business Model

Grameen Sanchar Society (GRASSO) is a non-government organisation which has developed a self-employment partnership model to bring information technology enabled services to the rural population.[8] Its origins began with a small group of seven persons with US$900 and has now grown to a 155 person organization with 23 offices and creation of 8000 self-employed persons and 33700 members with a total revenue of US$3.75 m in Sep 2006. Grasso set up the Community Information & Services Centre (CISC) whose objective was to increase awareness and information and provide transaction based business services and resulted in the creation of multi-purpose kiosks. Grasso provided the infrastructure, training and a transaction account with UTI bank. The centre provides portal services such as utilities bill payment, computing services, banking, microcredit and rural money transfer services and further bill payment, information services, document services and grievance facilities. Developers and managers with wide experience in portal development with critical application development were employed to provide the information technology based services. The CISC application has been localized which runs on Unicode compliant Bengali software/fonts, which makes it easy to use and manage.

The process for money transfer consists of the following steps:

*   CISC owner funds his CISC-Bank account or cash Cards or ICICI smart based accounts
*   Consumer uses the CISC facilities to buy services and pay bill
*   CISC logs into system, checks validity and pays online through Bank Payment Gateway
*   Bank transmits funds from CISC account to GRASSO account.

Grasso partnered with Microsoft and Jadavpur University to develop the information technology solutions and agricultural content respectively. Microsoft created the rural computer awareness network through Microsoft ULP Program. Presently, there are 50 GRASSO – Microsoft CTLC registered and operating throughout West Bengal. Jadavpur University provided the content and training for Agri-business and management certification courses so that the farming community is able to diversify into more modern crop, storage and marketing management. The communication links are carried out through a broadband connectivity in 30 towns, wireless based network in 8 suburbs, WiFi based implementation in few locations, with the rest through RAS-Dial-Up Internet. The com-

[8]Sengupta, S. 2006. Innovation to Transformation-GRASSO model. Secretary & CEO, Grameen Sanchar Society, Kolkata. Presentation made at Digital Asia, 27–28 November 2006, Kaula Lampur, Malaysia; and www.grassoportal.com

munication network infrastructure is provided in figure 8.4 below. Plans are being designed to introduce fixed wireless through WiMAX in cooperation with Siemens. The introduction of IT enabled services has led to the elimination of delays in remittance of funds. Use of MIS system provides up-to-date status of funds flow on a daily basis and is audited by the bank. The revenue model provides for the following:

- For the CISC owners – making available services for payments of bills on-line, and providing IT based services
- For Grasso – a nominal percentage for all transactions
- For Service Providers – better collection rates as opposed to current methods
- Banks – high transactions in pool accounts

The expected outcome is to make a CISC owner independent and have more earning opportunities with GRASSO offering the owner facilities to market products and services, such as community entertainment, insurance, health products, solar systems, farm equipment, agri-business, small producers' exchange and other utility services and bill collection facilities for service providers like BSNL Insurance etc. With the creation of the base applications, other services such as governmental information, land records, birth and death registration, online bazaar, rural auction centre, rural online bidding etc. could be added.

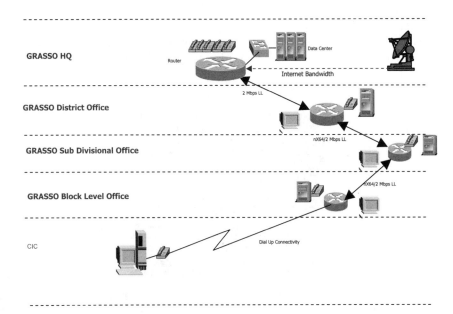

Figure 8.4   GRASSO communications network architecture

E-Choupal[9] - Transaction Based Rural Information and Knowledge Services Using Internet-India

E-choupal is a private sector led initiative to bring Internet-based applications to improve rural livelihoods for a majority of the population living in villages and remote areas of India. The program led by the Indian Tobacco Corporation was designed to help the farmers in the rural areas by bringing the latest price information prevailing in local and international markets of the farm products which are produced by the farmers; finding information about the latest agricultural technologies and practices; ordering agicultural inputs, such as seeds, fertilizers, pesticides, consumer products etc from ITC or its partners at prices lower than the local traders.

The program consisted of a computer linked to the Internet via phone lines and recently via VSAT connection which serves an average of 600 farmers in 10 villages covering a five kilometer area. Each E-Choupal costs between US$3000 and US$6000 for establishment, with an operating cost of about US$100 per annum. The computer is housed in the host farmer's residence. The farmer is called the *sanchalak* and receives training (see Figure 8.5). He serves as the main interlocutor between the computer and the illiterate farmers as well as takes a pledge to serve the community. The sanchalak gains prestige as well as a commission for all e-Choupal transactions. The information is provided free of charge to the farmers. Transactions are carried out online which are used as a basis to reward farmers for volume and value.

The e-Choupal initiative, which started as a small pilot initiative has proved to be so successful that it has been scaled up and now covers 31 000

Figure 8.5 Sanchalak training farmers in the use of computer in a village

[9]Choupal means gathering place in Hindi.

Figure 8.6 Smart cards are used for farmer identification

villages serving over 3.5 million farmers. Additional services have been added to this such as health, education, etc.

Lessons Learnt

The e-Choupal model demonstrated that a large private sector corporation can play a major role in increasing the efficiency of markets and agricultural systems benefiting farmers, rural communities and shareholders. It demonstrated the key role that information technology applications can play in improving rural livelihoods. Critical success factors were ITC's extensive knowledge of agriculture, retaining many aspects of the existing production system, including maintenance of local partners, the company's commitment to transparency and the respect and fairness with which both farmers and local partners are treated.

Internet in Rural Areas – the Example of the USA

Rural America has a greater share of older people than in urban and suburban areas – 43% of rural Americans are over 50 years of age and 18% are over 65 years as compared to 38% of non-rural Americans over the age of 50 years with 16% over 65 years. Older Americans go online at lower rates than other age groups. They have lower incomes than those living in urban and suburban areas – 33% of the rural population live in households with income levels less than $30 000 annually compared with 24% in non-rural areas. They are less educated than those in urban and suburban areas – 18% of rural Americans have college or higher degrees as compared to 29% in non-rural areas.

Access to the Internet by the population living in rural areas[10] in the USA varies significantly according to their location, income levels, level of education and ethnicity. Only 24% of rural Americans had high speed internet connections at home compared with 39% of adult Americans living in non-rural areas (see table 8.3). Internet penetration rate in rural areas in USA lagged the rest of the country by 8% at the end of 2005.

People living in rural areas use Internet more than those living in urban areas for taking classes for credit, computer games and sports, according to a survey conducted by Pew Internet research. The survey also found that being a rural internet user decreases the chances of using the Internet for making reservations for travel, banking online, reading classified advertisements and blogs (see summary of survey results in Table 8.4).

The margin of error is plus/minus 3% for Internet users in March, May and September surveys and plus/minus 2% for December surveys.

Table 8.3 Portrait of Internet access in the USA (in percent)[11]

Characteristics	Rural adults	Urban and Suburban adults
Home Broadband	24	39
Home Dial-up	29	21
Work	5	5
Other	3	3
Don't know	1	2
Non-internet user	38	30

[10] Note. The US Census Bureau classifies all areas as urban or rural. 'Urban' is defined as all territory, population, and housing units located within an urbanized area (UA) or an urban cluster (UC). It delineates UA and UC boundaries to encompass densely settled territory, which consists of:

– core census block groups or blocks that have a population density of at least 1000 people per square mile and
– surrounding census blocks that have an overall density of at least 500 people per square mile
  In addition, under certain conditions, less densely settled territory may be part of each UA or UC.

'Rural' consists of all territory, population, and housing units located outside of UAs and UCs. It contains both place and nonplace territory. Geographic entities, such as census tracts, counties, metropolitan areas, and the territory outside metropolitan areas, often are 'split' between urban and rural territory, and the population and housing units they contain often are partly classified as urban and partly classified as rural.
Source: http://www.census.gov/geo/www/ua/ua_2k.html last accessed on 12 September 2007.
[11] 2007. Pewinternet. www.pewinternet.org/pdfs/PIP_Rural_Broadband last accessed on 18 July 2007.

Table 8.4   Use of the Internet by rural population (in percent)

Uses	Rural Internet users	Urban and Suburban Internet users
A. When being a rural user decreases the chances of going online and using the Internet		
1. Buy or make a reservation for travel service/1	51	65
2. Bank online/2	34	43
3. Online classifieds/3	30	37
4. Read a blog/4	21	28
B. When being a rural user increases the chances of going online and using the Internet		
1. Download screensavers/5	28	22
2. Download computer games/6	25	20
3. Class for credit/7	15	11
4. Fantasy sports/8	9	7

Notes:
1. Denotes the Project's September 2005 March 2005 survey of 2251 adults (consisting of 1577 Internet users)
2. Denotes the Project's September 2005 March 2005 survey of 2201 adults (consisting of 1450 Internet users)
3. Denotes the Project's September 2005 March 2005 survey of 2251 adults (consisting of 1577 Internet users)
4. Denotes the Project's September 2005 March 2005 survey of 2201 adults (consisting of 1450 Internet users)
5. Denotes the Project's May 2005 March 2005 survey of 2001 adults (consisting of 1336 Internet users)
6. Denotes the Project's May 2005 March 2005 survey of 2001 adults (consisting of 1336 Internet users)
7. Denotes the Project's December 2005 March 2005 survey of 3011 adults (consisting of 1931 Internet users)
8. Denotes the Project's September 2005 March 2005 survey of 2201 adults (consisting of 1450 Internet users)

TREND TOWARDS BRIDGING THE RURAL-URBAN INTERNET GAP

There is a trend towards narrowing the gap between rural and non-rural areas in home broadband adoption in the United States. In this regard, the United States Department of Agriculture (USDA) initiated the Rural Development Broadband Program.[12] The goal of the programme is to bring affordable broadband to all rural Americans, which has been set by the President and the Congress of the United States Government.

[12] United States Department of Agriculture. May 2007. Rural Development: Bringing Broadband to Rural America. USDA, Washington DC.

Broadband has been considered to be a transformative technology which enables the rural communities to enhance their education and skills, provide access to regional, national and international markets. The rural broadband programme started as a pilot in 2000 with a $100 million loan program, which provided funds to rural communities with up to 20 000 residents. By the second year, a total of 28 loans in 20 states were made in the amount of $180 million. Since the pilot programme did not include any risk mitigating measures with regard to equity and loan security, about 30% of the loans defaulted, which are being restructured. Lessons learnt from this program have been incorporated in the expansion of the programmes. As of 2007, the programme has grown and has approved 70 loans in 40 states totaling over $1.22 billion. The broadband programme serves 1263 rural communities with a total of 582 000 household subscribers. Lessons learnt from the pilot program and subsequent scaling up have led to modifications in the programme design, some of which are:[13]

- Limit funding in urban areas where a significant share of broadband services are being provided by incumbent providers
- Ensure residents in funded areas are provided with broadband access more quickly
- Clarify and streamline the equity and marketing survey requirements
- Increase the transparency of the application process and promote a better understanding of all application requirements
- Ensure that projects funded are responding to the increasing demand for bandwidth.

The following two examples show how broadband has made a positive impact on the lives of rural communities in Michigan and Kansas.

Air Advantage Programme in Rural Michigan

The Air Advantage programme was founded in 2002 to provide high speed wireless internet service to schools and the rural community in Michigan's 'thumb area'. Two broadband loans were made, amounting to $1.53 million along with two community connect grants amounting to $433 700. The programme financed 300 miles of wireless backbone covering 2500 square miles. One example of a direct beneficiary was the Columbia Township Library – where a one room school house in 1965 has been transformed into a state–of-the-art library with computer facilities in 2005 which is used by 350 users out of 605 residents per month. The pro-

[13] Dulski, A. Beijner, H. and Herbertsson, H. 2007. Rural Development: Bringing Broadband to Rural America. USDA, Washington DC. May 2007.

gramme has resulted in the development of a distance learning program with seven school districts, using wireless broadband for connectivity. This has allowed the sharing of teacher resources and provided the students with more educational opportunities. The equipment needed for this purpose was financed from the distance learning and telemedicine grant programme from the USDA. Foreign language and advanced placement courses were added as a direct result of this programme initiative which were available to all the Huron county residents – with one teacher in one district providing the course for students in multiple districts. In addition, the programme provided for more than 50 students in Huron County to take college level courses while attending high school saving them from driving about 60 miles to the nearest college.

Nex-Tech – Rural Broadband for Businesses in Kansas

This programme upgraded the slow-dial up connection used by businesses to high-speed broadband connectivity to rural Kansas. The programme provided two broadband loans amounting to $6.57 million to Nex-Tech, a subsidiary of Rural Telephone, a leader in the telecommunications industry to bring broadband, video and voice throughout rural Kansas. Broadband technology provided the incentive and the business environment for one of the employers, Osborne Industries, a company operating in Kansas for over 30 years, to retain its operations in the town in Osborne and to expand its operations and provide employment to 112 employees. This was a direct result of bringing in broadband connectivity to a rural area. This has enabled Osborne Industries to increase its operating efficiency by using web-based systems for ordering, servicing and accounting purposes. Availability of broadband also attracted the building of a $54 million ethanol plant employing 34 people in Phillipsburg.

BROADBAND FOR ALL – EUROPEAN COMMISSION INITIATIVE AND EMERGING TECHNOLOGIES

The European Commission has launched the Bridging the Broadband Gap initiative whose aim is to bring high speed broadband Internet to all Europeans – especially to the less developed European Union (EU) areas. In this regard, 3G is a key enabler for realising the vision of Broadband for All[14], whose full capability to service the rural areas has not been fully

[14] Ericsson. 2007. Rural WCDMA – Aiming for nationwide coverage with one network, one technology & one service offering. Ericsson Review, No.2. http://www.ericsson.com/ericsson/corpinfo/publications/review/2006_02/05.shtml

recognised. The experience of Telstra, which operates in Australia, however, has demonstrated that 3G networks can serve as commercially viable enablers of communication and broadband services, even in areas with extremely low population density. In recent years, technological breakthroughs have occurred through the Third Generation Partnership Project (3GPP) which has resulted in the WCDMA (Wideband Code Division Multiple Access[15]) standards ensuring interoperability between different 3G networks for third generation cellular systems. These 3G networks are being rolled out in various parts of the globe and support wideband services like high-speed Internet access, video and high quality image transmission with the same quality as the fixed networks. In WCDMA systems the CDMA air interface is combined with GSM based networks. WCDMA network constitutes a cost effective way of rapidly deploying residential broadband services. Third-generation services with mass market potential include mobile and video telephony, data services, such as SMS and MMS, advanced data applications, including music downloads to a mobile handset, mobile broadband targeting laptop users, fixed-wireless broadband (ADSL alternative), mobile small-screen TV streaming, fixed PSTN-equivalent telephony and government, health and educational services.

COMMUNITY MULTIMEDIA CENTRES MODEL

The community multimedia centre (CMC) is a community-based facility which provides both community radio broadcasting and telecentre services consisting of Internet services and other information and communication technologies. This initiative has been launched by UNESCO[16] and has proven to be an effective form of intervention for community development. Exchange of information and knowledge has been found to strengthen social inclusion, public participation, education and open new

[15] There are two different modes of operation possible for WCDMA. (1) Time Division Dulex (TDD). In this duplex method, uplink and downlink transmissions are carried over the same frequency band by using synchronized time intervals. Thus time slots in a physical channel are divided into transmission and reception part. (2) Frequency Division Duplex (FDD). The uplink and downlink transmissions employ two separated frequency bands for this duplex method. A pair of frequency bands with specified separation is assigned for a connection. Since different regions have different frequency allocation schemes, the capability to operate in either FDD or TDD mode allows for efficient utilization of the available spectrum.
[16] Creech, H., Berthe, O., Assubuji, A.P., Mansingh, I., and Anjelkovic, M. 2005. *Evaluation of UNESCO's Community Multimedia Centres. Final Report.* Prepared by International Institute for Sustainable Development for United Nations, Educational, Scientific and Cultural Organization. Report number: IOS/EVS/PI/54.

opportunities. UNESCO has piloted 39 CMCs during the period 2000 to 2006 in Latin America/Caribbean, Africa and South Asia. An evaluation was conducted by the International Institute for Sustainable Development (IISD) between April and October 2005. This included site visits to operating CMCs in Mali, Benin, Mozambique, Tanzania, India and Nepal, and telephone surveys. The following were found to be the major achievements of the CMCs:

- Access to information through the Internet and radio makes a contribution to improving the quality of life in poor communities
- Longer term benefits are being realised within individual communities which include the gradual removal of barriers to social inclusion, access to knowledge on improving health, increasing agricultural productivity, better natural resource management and increased rural livelihood opportunities which combined together make positive impacts on poverty reduction.

Key Success Factors

The evaluation identified the following key success factors:

- Build on existing facilities
- Presence of strong commitment from the community
- Good integration of radio and telecentre components
- Diversification of content which meet the community needs, including the promotion of local culture
- Access to tools and expertise developed by UNESCO and other organisations
- Need to diversify sources for obtaining funds and build capacity to approach local/national governments and international development organisations for this purpose.

Key Factors for Scaling Up

An enabling policy environment was considered as an essential factor for scaling up. More specifically, the following policy issues need to be addressed:

- Need for a stable and possibly subsidised charges for rural connectivity
- Reliable and affordable power supply
- Making licenses for community radio stations easily available
- Ensuring freedom of the press

- Develop e-government applications and use CMCs as delivery points
- Promote the use of CMCs for the delivery of agriculture, health and education services for the rural population.

OPPORTUNITIES FOR INTERNET BASED SERVICES

In this section we look at some of the main opportunities for Internet based services that exist in rural areas.

E-Government Services for Rural Areas – the Big Opportunity

- Many governments around the world are placing increasing importance in using the Internet to:
- Provide better service delivery to citizens
- Improve services to business
- Promote transparency and anticorruption
- Promote empowerment through information
- Increase efficient government purchasing of goods, works and services.

Most of the governments however are still in the early stages of development as depicted in figure 8.7 where the use of Internet is limited to the

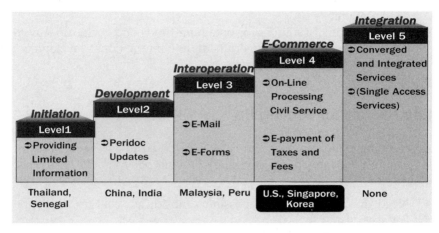

Figure 8.7 World trend of e-government[17]

[17] UN, Benchmarking e-Government, 2002.

provision of government information, with none of the countries in the world providing integrated services to the public. The problem is more acute in the rural areas and here is where the big opportunity lies.

Challenges

It is clear however that realisation of this opportunity requires finding new ways to address the following key challenges of e-government:

- Institutional challenges – to work in a collaborative manner at the policy and operational levels
- Culture – moving from manual to electronic ways of doing business by citizens
- Citizen's expectations that include:
- the impact of e-government on service delivery
- the necessity of having 'quick wins'
- inspiring trust in government information, services and processes
- Centralisation and Decentralisation – of key importance is decentralising sectoral services including management and delivery
- Cross sectoral e-government – cuts across all sectors, from agriculture to banking to commerce to debt management to economy to finance to governance to health to industry to jobs to security to trade, etc. The challenge is to establish which aspects to integrate and which cannot and should not be integrated
- Technology poses many real questions, such as do well-established technologies exist? And are these declining in costs with improved capabilities, capacity and efficiency?
- Training at all levels for providers and end users that focuses on increasing transparency and building public goodwill.

Another important element that requires discussion is the recovery of costs.

Cost Recovery

Experience from the provision of e-government Services in the USA indicates that cost recovery and ensuring sustainability of e-government services is a long-term process – to transition from a free to a fee based service. For example, a fee for services to the public was introduced about two to three years after the introduction of web-based government services as shown in figure 8.8. This period is likely to be much longer in developing countries. Creating long-term value is a long-term process spanning over two decades, as shown in Graph 2:

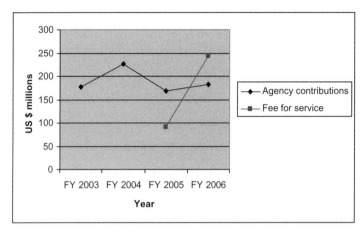

Figure 8.8 Financing e-government and cost recovery (example from USA[18])

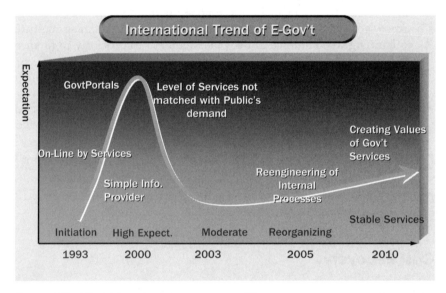

Figure 8.9 World trend of e-government[19]

[18] Office of Management and Budget (OMB). December 2005. Expanding E-Government Improved Service Delivery for the American People Using Information Technology. Executive office of the President. OMB. Washington DC 20503. http:// www.whitehouse.gov/omb/budintegration/expanding_egov_2005.pdf
[19] Gartner

Looking into the Future

With the increasing speed of and decreasing costs of Internet access, we hope to see the following in the years ahead:

- Integrated e-government strategies as an entry point to initiate multisectoral e-government operations
- Interoperability and common technology infrastructure across and within ministries
- Single window delivery of a range of e-government services that can be accessed:
  - in the comforts of the home
  - in information, knowledge and service centres
    - in rural and urban areas
    - interconnected within countries

Today, we are able to witness the *establishment of interconnected information, knowledge and business services centres* providing a range of e-government services

- tailored to meet the multisectoral and multidisciplinary information needs of the population
- providing free and fee based information, knowledge and business services
- using web-based services
- increased role of public sector to provide Internet connectivity to central nodes in rural areas to stimulate private sector involvement and to provide value-added demand driven services.

Using Universities

People living in rural areas have very limited access to the locally relevant, easy to understand, information and knowledge needed to improve their livelihoods and have low levels of literacy and affordability. On the supply side, there is unused potential and a human resource base in the higher educational institutions that have not been fully tapped for lack of incentives, resources and opportunities.

In this regard, universities could play a key role in assisting with the establishment and strengthening the operations of rural information, knowledge and services centres. Agricultural universities have access to science and research-based that their professional staff and students could convert into the kind of content that is both relevant and user-friendly for farmers and decision makers at local, national and regional levels.

Business Model

The basic design of the business model would be to use a committed network of agricultural universities to help establish sustainable rural information, knowledge and services centres (RIKS) linked and supported by each of the selected universities. These centres may either exist or be incubated by the university. The RIKS centres would serve as mechanisms for providing information, knowledge and a wide range of services to the rural population – ranging from free to fee based services to ensure sustainability over a period of time. The initial focus could be on meeting the information and knowledge needs of agriculture, but over time, a wide range of information, knowledge and services would be gradually developed to include several other sectors and disciplines based on the needs of various types of end users in the rural areas. The approach taken is consistent with the increasing trend to move from sector specific and individual information and knowledge services to the development of multisectoral and rural household oriented information and knowledge services. Recent advances in information and communication technologies, their declining costs and increased efficiency and speed make this possible.

E-Learning

Internet services have great potential in developing countries for distance learning. There is increasing demand for online education by working adults and for life long education. Distance learning and web-enabled education and training programs have the greatest potential to reach the remote and underserved regions. In China, for example, more than one million students study online and this number is increasing. Most of the online providers of higher education programs are based in the developed countries, but this is rapidly changing in developing countries, especially in Brazil, China, India, Turkey and Mexico. However, for online education to be effective and sustainable, it needs to combined with face-to-face instruction, built on a public-private partnership model. An approach to rural distance learning is provided in the following steps: (i) needs assessment – identification of learning needs of the target population, (ii) identification of course developers and content providers; (iii) developing learning modules; (iv) disseminating the learning modules through a variety of media – ranging from traditional (such as radio, television, telephone, etc) to modern information technologies (such as the internet, web-based technologies, etc.) and (iv) ongoing feedback and impact assessment for course improvement for different target learners.

ROLE OF MOBILE TECHNOLOGY FOR INTERNET DEVELOPMENT IN RURAL MARKETS

Mobile networks can serve rural consumers for about a tenth of the cost of a fixed network. And if you have a business case for building GSM coverage, you can offer Internet access at virtually no extra cost. This removes a great obstacle to bringing the Internet to all.

With more than 80% of the mobile world market and about 90% market share in developing markets, GSM has turned out to be the technology for the masses. Most GSM networks have now been upgraded, allowing operators to offer users not only traditional voice services but also medium-speed internet access and other data services. The availability of data access paired with mobility will drive information and communications technology (ICT) development and traditional as well as new data applications will have a mobile component for wider usage in the population.

GSM operators are rapidly rolling out cost effective transmission and radio networks in suburban and rural areas. As an example, population coverage in India was expanded from 30% to 60% in 2006 alone. In March 2007, the winners of an auction for the rights to create and run networks in remote rural areas in India were announced. Around the world, such networks are often subsidised by a universal service fund (USF) paid for by taxes on existing telecoms services. Auctions are held and the network operators that ask for the smallest subsidies win. They must then provide a certain number of public payphones and sign up subscribers. But something rather odd happened in India: in 38 of the 81 regions on offer, many mobile operators asked for no subsidies at all. India's biggest operator, Bharti Airtel, even offered to pay in 15 regions. As a result, barely a quarter of the INR40bn (US$920 m) available in subsidies is likely to be allocated. If operators reckon there is money to be made running mobile networks in some of the poorest parts of the world, have USFs had their day? Well, not exactly. Although Indian operators are rejecting subsidies for network equipment, they will still benefit indirectly from the fund because it is also used to subsidise the establishment of shared sites for mobile phone base stations. Even so, India joins other countries, such as Nigeria and South Africa, where commercial mobile networks are rapidly expanding into areas previously considered uneconomical. The advent of broadband Internet connections is also changing attitudes to USFs. Because broadband links can carry both voice and data, some countries are starting to subsidise broadband roll-out, instead of just concentrating on phones. And with so much of its fund unspent, India's government is drafting proposals to subsidise the provision of broadband to every village. Funds can also be used to provide high-speed access to schools, hospitals and local councils, and entrepreneurs can establish self-sustaining private community centres. This model is already taking hold in Uganda,

Mongolia, Burkina Faso and Malawi. For universal service funds and for telecoms in general, the trend is clear: phones first, broadband later.

From Voice to Data

Once GSM systems are in place, text messaging or SMS becomes available and can account for a significant part of communication. In markets with low-cost SMS, it is not unusual for the SMS volume to significantly out-number voice calls. In the Philippines, where sending and receiving SMS within an operator's network was initially free, SMS today contributes to more than 30% of operator revenues. Markets with high illiteracy and a multitude of local languages also have a high usage of SMS, primarily driven by the behavior of younger and more affluent consumer segments. It is important not to assume that advanced services are unsuited to emerging markets – leading-edge applications will almost certainly stim-ulate growth in such regions. While experience indicates that the bulk of revenues will come from basic voice services and peer-to-peer SMS, an attractive mix of value-added services can provide a significant additional revenue opportunity for operators. In particular, there is great interest in local data – news, community service information, prices, weather fore-casts and so on. Another interesting example is BubbleTalk, a 'click, talk and send' short voice-messaging service (VSMS or voice-SMS) initially offered by Malaysian operator DiGi. This 'talk and listen' messaging alternative to the 'type and read' service provided by SMS is much simpler to use than voice mail. Similarly, Indian mobile operator Bharti Airtel's Voice Portal on 646 service offers an interactive voice portal where cus-tomers can, for example, listen to sports news or download ringtones using voice commands. The portal logged more than 16 million minutes of use in its first 10 months of operation.

Easy to Add Data Services

While many basic data communications applications can be realised over SMS, many institutions – such as health authorities, educational institu-tions and government – will almost certainly need more bandwidth for mobile data. The relatively low investment cost of implementing GPRS and EDGE gives GSM operators a golden opportunity to offer subscribers mobile data services at a reasonable price.

Once GSM coverage is in place, it is easy to add a packet data service to the traditional voice service. GPRS, which handles packet-switched data and 'always on' functionality, has been around for more than 10 years and has evolved into an almost mandatory part of a cellular network. The additional cost required for the operator to offer data services is small. It doesn't necessarily require huge investments in a service network or a state-of-the-art advanced billing system. The service can be set up in a simple and robust way, with flat fee charging and focus on user friendliness. It doesn't have to be more complicated than that. Operators often have only a vague idea about the cost of transferring megabytes in a GSM network, which can result in consumers being overcharged. Few people are willing to spend several dollars for surfing the internet or using other kinds of mobile data services, especially when the price is per-downloaded data volume, something that they can't control. This is certainly the main reason behind the relatively slow uptake of mobile data usage in the developed world. In reality, an operator can offer data services over GSM for a small cost. How small depends on what quality of service it is guaranteeing. If only 'best effort' kinds of services are offered, then the cost of providing data is minimal. Voice traffic always has priority over data, and new, IP-based technology used in backhaul reduces the cost for data transmission, something perceived as a major obstacle in the past. The result is that if an operator has a business case for building GSM coverage, then it can also offer Internet access at next to no extra cost. No need for new sites, new spectrum or new terminals.

GSM is the Only Viable Choice Today

The telecom industry is debating the pros and cons of HSPA versus WiMAX, and which of the technologies is best suited to meet the future demand for mobile broadband services. It is important not to mix this debate with the one about how to bridge the digital divide – how to let the general population in developing markets get access to the internet.

In Western Europe, people normally mean a downlink speed exceeding 500 Kbps when they talk about broadband, often relating to a fixed-ADSL connection in the home. It is so easy to forget that, only a few years ago, many did not have any alternatives to dial-up connections over a normal telephone line.

Case Study: *Project MTN@Access*

The project MTN@ccess was launched in 2006 based on the belief that Internet access was going to be the next wave of mobility. South Africa already had more than 20 million mobile subscribers, but MTN didn't want to wait another 10 years to reach the same number of data subscribers, so the operator introduced a number of incentives to create demand. The project is a test to see whether a commercial model can be created by entrepreneurs to successfully run an internet cafe in a poorer community. Around that are several support infrastructures, such as financing and basic training. Then there's a portal where customers can get ready access to websites. If they're not familiar with the complexities of the Internet, they can easily find what they need through the portal, which constantly tracks their behavior and tries to bring the right content to them.

A key aspect is the simplicity of the model to the consumer; if they spend 10 minutes on the Internet, they understand exactly what it will cost them. Another aspect is the simplicity of the model to the entrepreneur – that they fully understand it is a profitable business. There have been some very encouraging moves – some of the early entrepreneurs, for example, set up business near schools and those cafés did very well. The ones who stood back and waited for things to happen lost out. This is the secret recipe – it is not what the operator does; they are just creating the fabric for an entrepreneur's success gave one tenth the speed of broadband. But for a majority of people in the developed world, this is still what they use to make their daily contact with the Internet. This situation is now changing quickly as the necessary factors come together to make deployment of mobile telephony based on GSM feasible. The biggest obstacle for this development has been the lack of availability of cheap telephones, but this has recently been addressed by the industry initiatives to develop ultra-low-end GSM terminals and there are also quite a lot of secondhand terminals flooding the market.

GSM equipment prices have fallen drastically over the past few years. Scale advantages, in combination with new mobile network features, have reduced the total cost of ownership for service providers, and this evolution has made mobile services affordable to the low-income segment. Suddenly, deployment cost is not the biggest issue.

In developing markets it is important to lower entry barriers for usage and offer services that improve quality of life. Because of the high cost of mobile phones, users commonly share phones. This sharing is often an ad hoc arrangement, with subscribers renting their phone for a call. In addition, operators are providing special community information centres. GSM operators also have the opportunity to be the first mover in offering internet services for all, holding a time-to-market advantage over any competing technology.

By equipping community centre PCs with data cards or embedded data modules, which today normally support EDGE as well as HSPA, operators can provide cost effective access to data services for everyone now as well as in the near future when bandwidth requirements increase and 3G licenses are released. HSPA can be introduced either into the entire network, as Telstra has done in Australia, or gradually, as MTN has done in South Africa with its MTN@ccess programme. The

MTN model, in particular, allows for data and voice to be carried over the GSM network when the subscriber is outside HSPA coverage. This approach enables the smooth, cost-effective introduction of HSPA.

True Mobile Broadband Services Over EDGE

When it comes to high bandwidth, the question that remains is does one need HSPA or WiMAX? Only the consumer can answer that. But for many markets, where people might spend hours getting water, food or traveling to work, it is hard to see a strong demand for high-speed data, mobility, roaming or other features that those in the developed world take for granted.

Coming to a Rural Area Near You

Access to Internet-based services is now readily available to people living in rural areas of Bangladesh. Grameen Phone, the country's largest phone company, has set up community information centres (CICs) in rural areas to make shared voice, Internet and data services available to those at the bottom of the economic pyramid. Some of the services now at their fingertips include video-conferencing with relatives and access to a wide range of information on such things as health, job opportunities, market prices for agricultural produce and government services. Piloted in February 2006, the CICs provide access to the Internet and other information-based services through Grameen Phone's nationwide EDGE connectivity. The CICs are located in rural areas, where the nearest point of access to the Internet is about 20–30 km away. There are more than 500 CICs in operation around the country. Each centre serves a community of about 40 000 people. The CICs are franchised, independent businesses run by a local entrepreneur. Each centre is equipped, at the least, with a computer, printer, scanner, webcam and EDGE modem to connect to the Internet. With expected revenue of US$6–7 per day and costing around US$1000 to establish, the centres are financially viable in about a year.

An affiliate of Telenor AS of Norway, Grameen Phone had more than 11 million mobile phone subscribers as of January 2007. In cooperation with Grameen Bank, the Nobel-Peace-prize-winning microcredit pioneer, Grameen Phone earlier introduced the internationally acclaimed Village Phone Programme, providing universal access to people in rural areas who normally could not afford to buy a regular subscription. It also enabled poor village women to earn a living by selling mobile phone services in their area to connect to the internet and run relatively simple applications for banking, trading or entertainment is a big leap in technology usage for many people in these markets. Most of the networks built during this century can easily handle EDGE data speeds. EDGE currently enables peak data rates of almost 300 kbps, and average data rates of around 120–160 kbps. With this bandwidth, almost all types of data applications available on mobile phones work well, and PC service is also feasible. EDGE data performance will be improved even more as EDGE Evolution is introduced in coming years. With EDGE Evolution, which is enabled by a software upgrade, bit rates can reach up to 1 Mbps and true mobile broadband performance can be offered over GSM.

Whatever strategy is used to address the broadband market, GSM with its evolution path to HSPA and beyond has a clear advantage over any alternative. Most of the investments, whether in the backbone or access networks, can be reused when new capabilities are introduced. GSM- and HSPA-originated traffic can use the same core network. Transmission networks can be reused and scaled to fit higher bandwidth requirements. Sites, in the core as well as the radio network, can be reused and new access technology can be introduced in parallel with what already exists. Even handsets and separate data modules, embedded in PCs as well as in data cards, will support multiple technologies. Provided the spectrum is available, the decision to evolve the network with new capabilities will be less a technical issue than a business issue. It will allow for gradual investments, ensuring profitability for the service provider while providing affordable services for all.

CONCLUSION

In addition to covering rural areas of developing nations like Africa, India etc., the definition of developing markets can also cover an emerging global economy such as China. Hundreds of millions of people in China live in rural areas and have not seen much of the country's economic gains. As part of its social responsibility policy, leading mobile operator China Mobile – with a nationwide GSM network serving 300 million subscribers – actively targets citizens living in rural areas, not necessarily with a profitable business case in mind. In Guangdong, the major province in southern China, China Mobile runs a successful EDGE data network serving half a million people. With attractive flat-rate data packages, sometimes bundled with voice services, and a generously dimensioned network, it handles millions of gigabytes of data every hour. The desire for mobile Internet is extremely strong in these markets; it is opening a whole world of opportunities and making people feel part of the all-communicating world.

9

Making It Happen: Enabling Communication in Developing Economies

OVERVIEW

At the beginning of the 20th century, the world was on the cusp of major social change as the result of emerging technology. No fewer than 500 companies had formed to produce this technology and early adopters had already begun to embrace the new invention, but it took the vision of one man to see the market potential for it. This man was Henry Ford and the technology was the automobile. Ford is reported to have repeatedly commented that there were many more poor people than wealthy people in the world. Ford's innovative contributions led to the lowering of technology costs, so that it was within easy reach of the mass market. In so doing, the automobile changed from an expensive toy for the wealthy to an indispensable cog in the production line of nearly everything produced today, including leisure.

Now, at the beginning of the 21st century, history appears to be repeating itself. Communication technologies mainly in GSM have changed, from a way for rich people to communicate, to essential factors in the production of almost every good or service. It is generally agreed that mobile communications is a the primary factor in promoting economic growth and wellbeing in developing nations irrespective if they are highly populous countries like India, China, Nigeria, Sudan etc. or smaller countries on the African continent.

Many of the less-developed corners of the world have embraced these emerging technologies and, in so doing, have become important players

Business Models for Sustainable Telecoms Growth in Developing Economies
S. Kaul, F. Ali, S. Janakiram and B. Wattenström
© 2008 S. Kaul, F. Ali, S. Janakiram and B. Wattenström

in the global supply chain, but sub-Saharan Africa has mostly been a notable exception to this phenomenon. This chapter will explore the policy and financial preconditions necessary to facilitate the rollout of communications infrastructure in developing nations. It has been said that the main challenge to the popularisation of the Internet is in fact neither illiteracy nor cost, but lack of relevant content. Without such content, the benefits of the Internet remain unclear for the majority of the population.

In addition to the paucity of content available in local languages, the price of access – whether to IP networks or GSM networks – remains a major constraint on growing to scale. Without regional backbones and Internet exchange points, international tariffs on traffic will continue to remain high. Fortunately, there are signs of progress in this regard. The Infinity Worldwide Telecommunications Group of Companies (IWTGC) has announced plans to install a second fiber-optic cable along the west coast of Africa, to compete directly with the current SAT-3 (South Atlantic Telecommunications Cable No. 3 transcontinental cable. The new cable, dubbed Project West Africa, hopes to sell bandwidth directly to service providers, thus lowering the cost of entry for service providers who, in order to gain access to the SAT-3 cable, must receive the unanimous consent of the entire consortium.[1]

The Project West Africa cable, expected to cost US$500m, will be financed entirely by the private sector. According to Russell Southwood writing on his website Balancing Act, the international traffic level has been suppressed by high pricing for international calls and for Internet bandwidth. Two main constraints have led to the suppression of traffic growth: monopolies over international gateways, and limited supply levels of fiber and satellite transmission infrastructure. The ending of incumbent exclusivity over international traffic is quickly coming to an end in many countries within the sub-Saharan Africa region and communications deregulation is opening opportunities for new international gateway licenses for Second Network Operators (SNOs) as alternative international operators and mobile operators.[2]

POLICY AND REGULATORY ENVIRONMENT

An effective policy and regulatory environment is critical for the deployment of communications services in developing countries, particularly as technologies converge. Clearly, ICT integration will lead to an increase in

[1] The Infinity Worldwide Telecommunications Group of Companies. 2007. www.iwtgc.com/images/balancingact268 last accessed on 11 July 2007.
[2] The Infinity Worldwide Telecommunications Group of Companies. 2007. www.iwtgc.com/images/balancingact268 last accessed on 11 July 2007.

competition that will ensure a reduction in transaction costs thereby enabling firms to make full use of economies of scale and to enhance regional infrastructure. Further it will encourage foreign direct investment (FDI) within the communications sector itself.[3] In addition, the search for an ideal regulatory environment had become more pressing given the growing digital divide between rural and urban Africa. The following are characteristics of an effective policy and regulatory environment:

Set Up an Independent Regulatory Authority

A well-regulated market, emphasises a number of factors such as transparency, accountability, due process and secures property rights. Regulatory agencies should be independent of the industry to be regulated and should possess the power necessary to force incumbents to allow access to their networks through reasonable interconnection fees. Without a functioning regulatory agency to enforce contracts, investors will continue to be reluctant to invest in foreign infrastructure.

Develop National ICT Policies that Encourage Competition

Historically, telecommunications markets in developing countries have consisted of a state-controlled monopoly. However this situation is quickly changing and is being driven by the convergence of communications technologies such as VoIP and WiFi. Furthermore, the majority of developing countries are beginning to appreciate the benefits of competition within this sector, but the attraction of economic rents accruing to governments has slowed the process. A second-order consequence of emerging technologies such as VoIP is the trend toward unified licensing regimes and away from discrete licenses for fixed, mobile and IP services. We address this topic in more detail later in this chapter.

In the past, regulation commonly took the form of capping tariff prices and telecom operator return on investment (ROI), but future regulatory emphasis should focus on improving access to networks. The use of Mobile Virtual Network Operators (MVNOs) should be encouraged. According to a study carried out by Pyramid Research (a telecoms consulting company), MVNOs provide a superior way to further communications growth in developing economies. The report goes on to point out that in imperfect markets such as those found in developing countries like sub-Saharan Africa (SSA), network operators have little incentive to open their networks to MVNOs, very often opting to hoard network capacity

[3] Tuju, R. 2007. Kenya's Information and Communication Minister, speech.

to accommodate future demand. It becomes clear that only when operators are forced to assume the full cost of their infrastructure are they
motivated to exploit unused capacity. To provide a viable business model,
MVNOs should be more cost-efficient than the network operator, while
at the same time offering a differentiated service, or extending the market
into areas the operator cannot or will not enter, such as rural areas.[4]

Prioritising Universal Access

Universal access to ICT – mainly telephony – is by far of much greater
concern only when communication markets are opened to the forces of
pure competition. Operators in competitive markets will generally behave
rationally by first connecting high density, largely urban, richer customers, while ignoring rural, poorer customers. For this reason, regulatory
agencies are requires to establish communication market environments
where operators will direct resources towards the development of communications that serve rural markets. Several methods of offering incentives to provide universal service access have been suggested, including
specific licenses for rural areas, as tried in South Africa, and the so called
'perishable' licenses which mandate that if operators have not provided
access to a particular region in a given period, then licenses for these areas
will be resold, perhaps through an auction.

Governments have successfully and efficiently induced network roll-
out via 'reverse auctions,' whereby operators compete to accept the lowest
possible subsidy to acquire the license. The South African mobile operator,
Vodacom has strongly protested the latest set of competition framework
changes in Africa, stating that these were most likely to be unfriendly
towards investors. In response, Vodacom has argued for a phased and
managed approach with no quick changes, thereby allowing new markets
to develop gradually.[5] Vodacom's view is that price regulation should not
be seen as the metric of successful regulation. Generally, most regulators
are in favour of driving prices down by addressing interconnection fees.
Instead, Vodacom's stance is that it makes sense to mandate open access
to more mobile operators (in the form of MVNOs) and allow competition
run its course. Any universal access strategy should include all the following key components:

Core Reform Programme

- Market liberalisation
- Regulation

[4] Zibi, G. 2007. Pyramid Research Report.
[5] Zibi, G. and Aytar, O. 2006. *Rethinking MVNO and MVNE economics: the future of mobile virtual models*. Pyramid Research.

- Capacity-building
- Privatisation
- Postal sector reform

Addressing Market Failures

- Rural access
- National backbone
- Post-conflict countries

ICT for Development Applications

- E-commerce
- E-government
- Civil society applications

There is strong evidence to support the notion that liberalised markets based on pro-competitive policies and regulatory frameworks provide strong high measure of support for improved access. In addressing market failures, mainly regarding the issue of rural access or the formation of a national backbone, it is clear that some public sector financing may be necessary. However, it is important to note that public support does not distort competition in growing ICT sectors – policy and regulatory interventions that can influence market development should be explored any form of public financing solutions are instituted. For instance, the introduction of output-based aid (OBA) schemes for rural access or backbone infrastructure development in less reformed environments is a risky proposition, unless it is preceded by detailed impact analysis to avoid subsidising what could otherwise be commercially viable operations if the regulatory environment was set up correctly.

Open the Spectrum for Emerging Technologies

The rapidly-changing business environment in which technology providers operate highlights the policy implications of the trend toward convergence, particularly in a country with a Universal Access policy. Because technology changes so rapidly, convergence dictates that regulation be technology neutral in order to avoid the stifling of innovation. However, technology neutrality is impossible in the absence of corresponding service neutrality. Service-limiting licenses are *de facto* technology-limiting. (The one caveat to this statement arises if competitors are not, in fact, perfect substitutes for each other, as in the case of an infrastructure-owning telephone incumbent and a mobile virtual network operator. The recognised need for these two attributes – technology neutrality and service neutrality – has given rise to so-called converged licenses. Converged licenses recognise that the service offerings of operators are increasingly becoming substitutes, and, in fact, operate using the same finite resource, the frequency spectrum. Such licensing schemes bring the entire spectrum under the jurisdiction of one regulatory agency. Spectrum licensing schemes offer several advantages. First, by bringing the entire usable spectrum under one regulatory body, issues of radio interference can be addressed quickly and easily. Second, operators will have clear property rights to portions of the spectrum. Third, single regulatory agencies frequently permit arbitrage among license-holders, enabling spectrum-consolidating and wealth-creating transactions. Such consolidations may lead to economies of scale, as technology providers (both hardware and software) are able to take advantage of greater market reach. Fourth, a single regulatory agency can be more efficient in negotiating conflicts at the periphery of a regulated environment, such as with neighboring countries. Finally, with a single, unified license based on spectrum frequency instead of technology or service, and the ability to resell unused spectrum, incentives exist to utilise existing infrastructure efficiently, since unused capacity carries an opportunity cost. This has profound implications for rolling out infrastructure to rural and other currently underserved areas.

In countries such as Mali, the incumbent telephone operator has little incentive to resell its unused capacity, and may, in fact, be justifying hoarding capacity, based on anticipated future demand for its services. While future needs may be a legitimate reason to reserve capacity, this penalises potential customers who would be willing to pay for services now, and there may be MVNOs willing to service this low-ARPU market. In a regulatory environment where spectrum reselling is permitted, predatory or spiteful hoarding is reduced. The Ugandan communications market has often been highlighted as a very favorable regulatory environment. Local operators agree that telecommunications regulation there is much more progressive than then anywhere else, that the Uganda Communications Commission is fair and equitable and that contracts and agreements are generally honored. Generally, there is a feeling among operators that they can concentrate on growing their businesses instead of wasting time sorting out bureaucratic problems.

Enabling Capital Access

A basic goal of regulatory policy should be to enable resources to go to their highest-valued use. Currently, many SSA countries have restrictions on the percentage of foreign ownership in telecom operators. Such restrictions only hinder infrastructure roll-out, as it shrinks the available capital. Similarly, taxing handsets as luxury items constrains end user access to affordable hardware.

Public-private partnerships, through such organisations as the International Finance Corporation (IFC) and the CDC group (formerly the Commonwealth Development Corporation), have had some success in stimulating infrastructure roll-out directly by providing access to capital, but the record is decidedly mixed. It has been stated by some telecom operators that development assistance is at best a waste of time and at worst an active impediment. Uganda Telecom, for example, has inherited a patchwork of incompatible legacy systems, all built with donor assistance tied to the use of particular suppliers, much of which is now fit for nothing but scrap. None of the highly publicised regional government- and donor-led infrastructure projects appears to be making anything more than minor blips on the radar screen. According to one Tanzanian operator, as much as 75% of assisted projects fail to work. Clearly then, should a project be commercially viable, then it should be done commercially.

Grazing the Barriers of Credit for Low-Income Groups

One successful example of a public-private partnership (PPP) is in the creation of microcredit initiatives, as in the Grameen Village Phone. Alternatively, donor organisations could provide funding specifically for rural and unserved areas, using an output-based Aid (OBA) approach.

Building Human Capacity

Several project managers[6] were of the opinion that the training in ICTs at university level and the difficulty of finding qualified applicants to service networks and to create content was a major constraint for ICT development in developing countries. Additionally, low literacy rates not only impede the uptake of ICT services on the demand side, but also hold back the roll-out of infrastructure and the development of innovative content on the supply side. Addressing issues of literacy and education are beyond the scope of this paper, but nonetheless are worth mentioning. One way in which the private sector, particularly communications operators, could address the issue of human capacity is to adopt a vertical integration approach. Operators need to leverage their marketing abilities to promote value-added services on their networks. There have been many successful

[6] Various formal and informal discussions held with multiple communications project managers across Africa, Middle East and Europe during 2006–2007.

examples of this in the United States, where communications operators sell handsets with value-added services such as AOL's Instant Messenger already installed. Sonatel could try a similar approach with Manobi's value-added service in Senegal. PPPs can also be beneficial in improving human capacity through investment in training programs at university level in order to develop demand for ICT services, as well as in the workforce to support such technologies. Such an approach should also trigger innovation and content creation on the supply side, creating a virtuous cycle in the manner. Donor organisations could try to address inequalities in access to ICTs between genders through training programs focused directly at females. Finally, donor organisations, whether through PPPs or alone, can serve as business incubators for value-added service providers.

Infrastructure Deployment in Rural Areas of Developing Economies Can Be Profitable

In chapter 8, two illustrative case studies (GSM and GSM/WiMAX) were presented supporting this claim. Although payback periods will be longer in rural markets than in urban markets, it is generally understood that the demand for ICT is strong among the rural poor of developing countries.

Input from Stakeholders at Every Step of the Value Chain is Essential

Infrastructure rollout within rural areas of developing countries will largely be demand-driven. Communications network operators need input from their customers regarding what services they are willing and able to pay for. In turn, communications operators and service providers must pay careful attention to the needs of end users. Local experiments and pilot projects financed through PPP's can seriously prove there is an unmet demand in these areas. Without the necessary buy-in from end users, network roll-out based solely on technological concerns has no guarantee of success. Communications operators must ensure they are offering infrastructure in the quantities and qualities consumers want and can afford. In other words, they must ensure that there products and services are customised to end user requirements. When Manobi lobbied Sonatel to build a tower in Kayar, for example, the result was a larger mobile footprint for all consumers and a large number of subscribers added to the network.[7]

[7] Ericsson. 2007.

Mature and New Technologies Offer a Chance to Overcome the Lack of Legacy Infrastructure in Rural Areas

Mobile technology offers a chance for developing economies to deploy networks that overcome the lack of legacy fixed telephone lines. The challenge of ensuring universal access to not only telephone services but also to the Internet can benefit from new technological developments. For example, technological evolutions such as packet radio offer reach and capacity in various situations (from dense urban to very low-density rural areas) while reducing the Total Cost of Ownership (TCO) for operators. Additionally, network convergence through introduction of IP packet technology in the core, and eventually access, portion of networks will open the way for fixed mobile convergence and IP telephony.

Financial Innovation Will Allow Operators to Deploy More Aggressively into Rural Areas

This book has highlighted several innovations that allow communications operators to obtain sufficient financing to cover the cost of infrastructure deployment. For instance, a pay-as-you-grow scheme allows an operator to pay for network capacity only as demand warrants and to avoid paying for unused capacity. Under-Serviced Area Licenses (USAL) and Output-Based Aid (OBA) schemes can also sustain such inroads by operators into rural areas. In addition to USALs and OBA, policymakers and regulators can do much to facilitate further infrastructure rollout among the rural poor. Below are some issues for policy-makers to consider:

Encourage Market Entry by a Full Range of Operators, Including Large-Scale and Micro-Entrepreneurs

Spectrum licensing and the reselling of unused spectrum can promote the efficient use of existing infrastructure and should stimulate competition among operators, by granting micro-entrepreneurs access to networks without the need for infrastructure investment. Impediments such as prohibitions on foreign ownership should be removed, to encourage investment in infrastructure as well as in service providers.

Encourage Public-Private Partnerships to Create an Enabling Environment

These efforts should include training for end users in the use of ICT-enabled services, and the marketing of value added services on existing networks. For example, REOnet (a consulting company based in Mali) is providing training to individuals in the use of ICTs for telemedicine, as well as programming in a Linux environment. Additional funding for training would stimulate demand for services similar to Project IKON (an experiment in tele-radiology). Access to financing for service providers is also a critical issue that has to be addressed by policy makers. Private sector investment has an important role to play in ICT development, and all sources of funding must be considered, including those offered by public-private partnerships, in order to lower the cost of small business loans for ICT-related services, and increase access to private equity.

Ensure Technology- and Service-Neutrality of Regulatory Policies

The dizzying pace of change in information and communication technologies virtually ensures that regulations that are technology or service-specific will become out of date very quickly. Technological evolution in the telecom sector will continue to bring further opportunities for developing countries. Regulatory framework flexibility should allow a wide diffusion of these emerging technologies.

Above All, Ensure that the Regulatory Environment is Transparent, Pro-Market, and Fair

Regulators should simplify existing licensing procedures to ease market entry and operations. By combining regulatory authority for telephony and IP networks, developing countries can foster innovative service offerings at competitive prices. Rural consumers will demand services that enrich lives and increase access to information in ways that only they can anticipate. By easing regulatory friction, regulators can hasten the day when even the rural poor in developing economies have access to information and communication technologies, and can contribute to solutions to the vexing issues outlined by the Millennium Development Goals. Children in Senegal and tuberculosis patients in South Africa are ready.

CONCLUSION

Policy and financial environments can have a profound influence on the spread of ICTs in developing economies, particularly in as-yet unserved rural regions. Regulatory policies should focus on encouraging investment, including foreign investment. With this goal in mind, regulation should be technology- and service-neutral, and should be characterised by transparency, clarity, fairness and flexibility in anticipation of future technological developments such as VOIP etc. Regulation should also seek to minimise the need for litigation. Above all, enabling environments should seek to foster competition. It is up to each country to find the balance between protecting the property rights of license holders and promoting innovation in ICTs. These issues apply to all countries, but the potential benefit of relaxed control over the use of the spectrum is greatest in underdeveloped countries such as those found in sub-Saharan Africa, Nigeria, China, Russia, Thailand, etc.

Index

Business Models for Sustainable Telecoms Growth in Developing Economies
S. Kaul, F. Ali, S. Janakiram and B. Wattenström
© 2008 S. Kaul, F. Ali, S. Janakiram and B. Wattenström